Fighting A Nuclear-Armed Regional Opponent: Is Victory Possible?

Contents

Fighting A Nuclear-Armed Regional Opponent: Is Victory Possible? ii

Publishing Information iii

AI-generated Bibliographic Keywords iii

Publisher's Notes iii

Abstracts iv
    TL;DR (one word) iv
    Explain It To Me Like I'm Five Years Old iv
    Synopsis iv
    Scientific Style v
    Action Items (Prospective) v

Excerpts vii
    Seven Most Striking Passages vii
    Grounds for Dissent ix
    Red Team Critique xi
    MAGA Perspective xiii

Fighting A Nuclear-Armed Regional Opponent: Is Victory Possible?

Nimble Books LLC: The AI Lab for Book-Lovers

Fred Zimmerman, Editor

Humans and AI making books richer, more diverse, and more surprising

Publishing Information

- (c) 2024 Nimble Books LLC
- ISBN: 978-1-60888-321-9

AI-generated Bibliographic Keywords

electromagnetic pulse book; electromagnetic pulse effects; nuclear weapons effects on systems; nuclear warfighting strategies; cold war nuclear stand-off; nuclear weapons physical mechanics; hemp effects on ground systems; preventative measures for emp; scaling laws nuclear weapons; nuclear conflict military operations; regional nuclear powers; tier three nuclear states; strategic operations nuclear forces; Japan's nuclear program; Soviet Union nuclear strategy; nuclear weapons delivery systems; limited nuclear use military policy; nuclear-armed regional powers; nuclear weapons and Cold War

Publisher's Notes

This annotated edition illustrates the capabilities of the AI Lab for Book-Lovers to add context and ease-of-use to manuscripts. It includes several types of abstracts, building from simplest to more complex: TLDR (one word), ELI5, TLDR (vanilla), Scientific Style, and Action Items; essays to increase viewpoint diversity, such as Grounds for Dissent, Red Team Critique, and MAGA Perspective; and Notable Passages and Nutshell Summaries for each page.

Abstracts

TL;DR (one word)

Nuclear.

Explain It To Me Like I'm Five Years Old

Alright, imagine you have a big, magical flashlight. When you turn it on, it sends out a super strong burst of light and energy. This burst is so powerful that it can make all the electric gadgets around it stop working, like your toys, TV, or even computers. This magical burst is called an Electromagnetic Pulse, or EMP for short.

Now, sometimes people talk about a really big EMP that can happen if a special kind of bomb, called a nuclear bomb, goes off high up in the sky. This big EMP can make lots of things on the ground stop working, even if they're far away from where the bomb exploded. It can mess up things like lights, phones, and even cars. People have to think of ways to protect these things from getting broken by the EMP.

To keep things safe, smart people have come up with different tricks. They might wrap important gadgets in special materials or build strong shields around them. This way, if an EMP happens, the gadgets inside these shields will still work. It's a bit like putting your favorite toy in a strong box so that it won't get broken if something bad happens.

Synopsis

This study, prepared by the Center for Strategic and Budgetary Assessments for the Office of Net Assessment in the Office of the Secretary of Defense, examines the increasing likelihood that the United States will face nuclear-armed regional adversaries in future conflicts and explores the implications of this likelihood for US strategy, military planning, and force structure. The report also examines how the history of nuclear proliferation suggests a new period of punctuated equilibria in which the number of states possessing nuclear weapons will increase. The study then analyzes how emerging nuclear powers may seek to take advantage of the apparent brittleness of current US strategic and operational responses to the nuclear threat, which are rooted in the binary choice between inaction and nuclear retaliation. The authors identify three basic strategic and operational responses available to the US: Option A, or the status quo, which relies on overwhelming conventional military superiority coupled to nuclear deterrence; Option B, or a Moderate Adaptation Strategy, in which the United States makes a major investment in counterforce and active/passive defense capabilities to neutralize a regional nuclear threat without recourse to a large-scale nuclear retaliatory response; and Option C, or an Aggressive Adaptation Strategy of Neutralization and Defeat, in which the United States invests in the full range of counterforce, active defense, and C4ISR capabilities to neutralize a regional nuclear threat and to prevail in a protracted regional conflict despite nuclear weapons use by the adversary. After analyzing both the requirements and feasibility of Options B and C, the study concludes with several recommendations to prepare the US for this challenge.These recommendations include developing new concepts of operations, training for counter-nuclear operations, operating from heavily defended bases within the theater of operation, preparing for dispersed combined arms operations, and determining an appropriate EMP tax to ensure that all future combat systems can function in an electronically disturbed environment. Finally, by preparing for a regional nuclear conflict, US military forces will also obtain a robust capacity to deal with opponents equipped with a large arsenal of precision-guided, tactical and theater range munitions.

Scientific Style

Abstract

The Cold War era saw the United States and the Soviet Union engaged in a nuclear stand-off, characterized by indirect warfare and covert operations. Amidst this tension, the emergence of nuclear weapons and their long-range delivery systems marked a significant Revolution in Military Affairs (RMA). This paper delves into the mechanics of Electromagnetic Pulse (EMP) phenomena, specifically High-altitude Electromagnetic Pulses (HEMP), and their effects on various systems. The study systematically explores the physical mechanics of EMP, its impacts on ground and other systems, and proposes preventative measures. Additionally, using scaling laws, the paper estimates the potential effects of nuclear weapons. The strategic context of nuclear warfare is discussed, emphasizing the importance of a robust investment policy that enables military operations even under limited nuclear exchanges, potentially involving states with smaller nuclear arsenals like North Korea or early-stage nuclear states such as the Soviet Union and China. The analysis highlights the strategic challenges posed by mature Tier Three nuclear powers like contemporary China, drawing parallels to Cold War dynamics. This study underscores the necessity of appropriate weaponry and strategic frameworks to navigate the complexities of nuclear conflict, referencing essential resources such as "Japan's Nuclear Program" and "Strategic Operations of the Nuclear Forces."

Action Items (Prospective)

- Reflect on the key concepts and information presented in the book. Consider how the information about electromagnetic pulses (EMP), their mechanics, and their effects on various systems could be relevant to your own interests or work.

- Consider the implications of EMPs and nuclear strategies in today's geopolitical environment. Think about how current international relations might be influenced by the concepts discussed in the book.

-Apply knowledge about preventative measures and scaling laws to real-world scenarios. This could involve assessing the vulnerability of current infrastructure to EMP attacks and identifying potential mitigation strategies.

- Discuss the book's content with peers or in a study group to gain different perspectives and insights.

- Stay informed about ongoing developments in nuclear technology and international security by following relevant news sources and publications.

- Use the book as a foundation for further research or projects related to EMPs, nuclear strategy, or national security. Consider writing papers, creating presentations, or conducting experiments based on the knowledge gained.

- If interested in policy-making or strategic planning, think about how the insights from the book could inform decisions and strategies at the governmental or organizational level.

- Revisit specific sections or chapters that were particularly complex or interesting to ensure a thorough understanding of the material.

Excerpts

Seven Most Striking Passages

Here are seven striking, important, memorable, or insightful passages from the book "Fighting a Nuclear-Armed Regional Opponent: Is Victory Possible?" by the Center for Strategic and Budgetary Assessments:

1. **"The bottom line is quite simple: the United States should have military and technological options to deal with emerging nuclear powers."** (Executive Summary)
    - This statement concisely encapsulates the core argument of the book: the US must adapt its military strategy to confront the growing threat of nuclear-armed regional powers.
2. **"Never fight the United States without nuclear weapons."** (Page 4)
    - This quote from the retired Chief of the Indian General Staff highlights the strategic dilemma facing the US. A nuclear-armed adversary may choose to escalate to a higher level of conflict to negate the US's conventional advantage.
3. **"A significant disjuncture exists, therefore, between the increasing likelihood that US adversaries may possess nuclear weapons and American lack of preparedness to fight them."** (Page 5)
    - This passage underscores the critical mismatch between the rising threat of nuclear proliferation and the US military's focus on conventional warfare and counterinsurgency operations.
4. **"There are severe and intractable difficulties in developing forces for nuclear warfare."** (Page 11)
    - The book acknowledges the immense challenges associated with preparing for a nuclear war. The demands of survivability and effectiveness in a nuclear environment are conflicting and difficult to reconcile.
5. **"The current IT-intensive RMA both eases and complicates the problems associated with operating in a nuclear battlespace."** (Page 12)
    - This passage explores the double-edged sword of the information technology-driven revolution in military affairs. While it offers advantages in precision and conventional warfare, it also makes the US more vulnerable to the disruptive effects of nuclear weapons.
6. **"A major decision for Option B will be whether to execute counterforce strikes in advance of decisive operations, with the timing and strategic objectives of the operations dependent upon the outcome of the counterforce campaign."** (Page 9)
    - This passage highlights the complex strategic choices facing the US when confronting a nuclear-armed adversary. The decision to strike an opponent's nuclear capabilities before or during a conflict is fraught with consequences and risks.

7. **"The challenge of defeating superhard and dispersed underground targets remains. This challenge raises the prospect that some deeply buried targets can only be attacked by a new generation of nuclear EPWs [Earth Penetrating Weapons]."** (Page 54)
    - This passage reveals the technical and moral complexities of countering a nuclear threat. The use of nuclear weapons to target hardened underground facilities raises significant ethical and strategic considerations.

The book's overarching recommendations call for a significant shift in US defense strategy, including:

- Investment in Counterforce Capabilities: The US needs to develop a more robust ability to locate, track, and destroy an adversary's nuclear weapons, including those hidden underground.
- Enhancement of Active Defenses: The US must invest in advanced missile defense systems, including boost-phase interception, to counter a broader range of missile threats.
- Resilience to EMP: All new military capabilities, including C4ISR and ground forces, should be hardened against the effects of electromagnetic pulse from nuclear detonations.
- Development of New Operational Concepts: The US must prepare for the possibility of conducting operations in a nuclear-shadowed environment, including the use of dispersed and mobile forces, and reliance on air-based logistics and fire support.
- Training for Nuclear Operations: The military needs to conduct more realistic exercises and training scenarios to prepare for the complexities of operating in a nuclear environment.

"Fighting a Nuclear-Armed Regional Opponent: Is Victory Possible?" argues that the US must move beyond a reliance on deterrence and a focus on conventional warfare and embrace a more adaptable and robust strategy to face the challenges of a proliferating nuclear world. The book's recommendations, while ambitious, represent a necessary step in ensuring US security in an increasingly complex and dangerous international landscape.

Viewpoints

These perspectives increase the reader's exposure to viewpoint diversity. No endorsement of any particular view is intended.

Grounds for Dissent

My dissenting view centers around the assumptions and conclusions drawn in the report regarding the utility and sustainability of military operations in a limited nuclear conflict. The report appears to underestimate the catastrophic and multi-faceted impacts of even limited nuclear engagements, which could render traditional military operations untenable and lead to severe humanitarian and environmental crises.

Firstly, the report suggests that the United States could conduct military operations in the face of limited nuclear use, drawing parallels to historical Cold War scenarios. However, this perspective fails to account for the significant advancements in nuclear and conventional weaponry, as well as the increased interconnectedness of global systems since the Cold War. Modern nuclear weapons are far more destructive, and their use—even in a limited capacity—could lead to widespread fallout, electromagnetic pulse (EMP) effects, and long-term environmental damage. These consequences would severely disrupt communication, transportation, and supply chains, making sustained military operations highly improbable.

Secondly, the report's discussion on the effects of High-altitude Electromagnetic Pulse (HEMP) on ground systems appears to be overly optimistic about the resilience and adaptability of these systems. EMP effects can cripple electronic infrastructure, which is crucial for modern military operations. The complexities of re-establishing operational capabilities post-EMP attack are significantly underplayed. According to a 2008 report by the Commission to Assess the Threat to the United States from EMP Attack, the U.S. is highly vulnerable to EMP attacks, which could lead to long-term power outages and critical infrastructure failures. This vulnerability indicates that the assumptions made in the report regarding the continuity of military operations are overly simplistic and potentially dangerous.

Furthermore, the geopolitical context has evolved substantially since the Cold War. The rise of cyber warfare, the proliferation of advanced conventional weapons, and the potential for asymmetric warfare tactics by state and non-state actors complicate the strategic landscape. Relying on strategies that hinge on Cold War-era dynamics does not adequately address these modern threats. For instance, cyber attacks can disable critical infrastructure without the need for kinetic action, and these attacks can be coordinated with or in lieu of nuclear strikes, further complicating military response.

Additionally, the environmental and humanitarian impacts of nuclear warfare, even on a limited scale, are grossly underestimated in the report. Studies, such as those published in the journal *Science Advances* (, demonstrate that even a limited nuclear exchange could lead to severe climate disruptions, agricultural collapse, and widespread famine. These effects would not be confined to the immediate area of conflict but would have global repercussions, destabilizing regions far from the initial blast zones and complicating humanitarian and military responses.

In conclusion, the report's assumptions about the feasibility and strategic value of conducting military operations amidst limited nuclear use do not adequately consider the full spectrum of modern threats and consequences. A more cautious and comprehensive approach is required, one that incorporates the latest technological, environmental, and geopolitical developments. The risks of underestimating the impacts of nuclear conflict are too great to rely on strategies rooted in outdated paradigms.

Red Team Critique

Red Team Plan:

The strategy outlined in the document reveals an emphasis on readiness and operational capability in the context of limited nuclear use. To counter this, our approach will exploit several single points of failure, asymmetric vulnerabilities, unsustainabilities, and political fragilities inherent in their strategy.

Single Points of Failure:

- **Command and Control Systems**: Focus on cyber-attacks targeting the central command and control infrastructure. Disruption of communication channels can lead to disarray in the execution of their nuclear strategy. Phishing campaigns, malware, and DDoS attacks can incapacitate central nodes and satellite systems.

- **Supply Chain Vulnerabilities**: Disrupt the supply chain for critical components of their nuclear delivery systems. This might include targeting manufacturers of key electronic components or the logistical chain that ensures these parts reach their destinations. This can be achieved through sabotage, insider threats, or cyber-attacks aimed at critical infrastructure.

Asymmetric Vulnerabilities:

- **Non-Nuclear Countermeasures**: Develop and deploy advanced electronic warfare capabilities to neutralize the effects of HEMP. For instance, deploying EMP-resistant drones to disrupt communication and surveillance systems could enhance our asymmetry.

- **Psychological Operations (PsyOps)**: Utilize misinformation and propaganda to create confusion and distrust within their nuclear command structure. This includes spreading false information about the integrity of their nuclear arsenal, leading to internal audits and delays.

Unsustainabilities:

- **Economic Strain**: Increase the financial burden on maintaining a nuclear arsenal by engaging in an arms race where we invest in cost-effective countermeasures like EMP hardening and advanced missile defense systems, forcing them to spend disproportionately more to overcome these defenses.

- **Technological Overload**: Encourage rapid technological change and obsolescence in their systems by continually introducing new counter-technologies. This can be achieved through continuous innovation in cyber warfare, electronic warfare, and missile defense systems that force them to constantly update and replace their equipment.

Political Fragilities:

- **Alliance Disruptions**: Target alliances and coalitions that support their nuclear strategy. Through diplomatic channels, covert operations, and intelligence sharing, drive wedges between allied nations to create distrust and political friction. For example, revealing or fabricating proof of espionage or betrayal could weaken these alliances.

- **Domestic Unrest**: Exploit internal political divisions and social unrest. Support dissident groups financially and ideologically to create a more chaotic and divided political landscape, which can undermine their ability to maintain a coherent nuclear strategy. This can include leveraging social media platforms to amplify existing grievances and instigate protests or other forms of civil disobedience.

Operational Tactics:

- **Simulated Attacks**: Conduct simulated cyber or kinetic attacks that mimic the effects of an EMP, compelling them to reveal their response protocols and vulnerabilities. This can include high-altitude detonations or cyber intrusions that mimic the effects of an EMP without causing real damage.

- **Deception Operations**: Use decoys and misinformation to mislead their intelligence about our actual capabilities and strategies. This can involve creating fake technology demonstrations, disseminating false documents, and planting misleading intelligence through double agents.

By focusing on these areas, we can create a multi-faceted approach that undermines their strategy from multiple directions, leveraging our strengths to exploit their weaknesses effectively. This not only disrupts their current capabilities but also forces them into a reactive posture, continually adjusting to our initiatives and thereby increasing their operational and strategic instability.

MAGA Perspective

In an increasingly polarized market for English-language books, it must be assumed that readers will often come into contact with views of the topic that are that deeply skeptical of conventional wisdom. Consider this section an inoculation.

This document is yet another example of the bureaucratic nonsense spewing out of our so-called 'defense experts' who have clearly lost touch with the real world. We know that this country is under threat from enemies both foreign and domestic, but instead of focusing on immediate, practical solutions to protect American interests, they drown us in technical jargon about electromagnetic pulses and scaling laws. The people behind this useless paperwork are either incompetent or deliberately sabotaging our nation's security by keeping us wrapped up in theoretical babble instead of gearing up for the real threats we face today.

Let's be real, folks. This obsession with nuclear nuances and the minutiae of long-range delivery systems is a distraction. It's almost as if these so-called experts are more interested in playing armchair generals than in defending our homeland. When they talk about "limited nuclear use" and the "second RMA," they completely ignore the fact that our adversaries aren't playing by any rules. North Korea, China, Iran—these are rogue states who aren't interested in limited anything. They are interested in wiping America off the map. And what do we get? A document that reads more like a science fiction novel than a strategic defense plan.

And while they're busy pontificating on nuclear warfighting policy, our borders are wide open and our cities are plagued by violence and drugs. It's almost as if the authors of this document have never set foot outside their ivory towers. They pontificate about nuclear deterrents while real Americans are struggling to keep their families safe from crime and economic hardship. Prioritizing complex nuclear strategies over basic homeland security is an absolute betrayal of American citizens who suffer daily from the real threats here at home.

This document also reeks of elitism. It's packed with references to academic studies, foreign policies, and international treaties that the average American couldn't care less about. Who cares about Japan's nuclear program or the strategic operations of some outdated Soviet study when our country is teetering on the edge? The working man and woman need action, not theoretical mumbo-jumbo. This document shows just how out of touch the establishment is with the common people who demand real, robust, and immediate defensive measures.

Lastly, let's not overlook the hidden agenda here. They talk about "appropriate weapons" and "strategic challenges," but what they really mean is lining the pockets of defense contractors with taxpayer dollars. This document is essentially a blank check for the military-industrial complex. They want to scare us into submission with complex terminology and far-fetched scenarios so they can justify their bloated budgets and pay for their pet projects. Meanwhile, Main

Street America gets left in the dust once again. It's a disgrace, and it's high time we demand accountability and a shift in focus towards practical, effective defense strategies that protect Americans now, not in some abstract future.

Fighting a Nuclear-Armed Regional Opponent:

Is Victory Possible?

by

Center for Strategic and Budgetary Assessments

Prepared for the Office of Net Assessment
Office of the Secretary of Defense

Contract No.: DASDW01-02-D-0014-0071

April 2008

The views, opinions, and/or findings contained in this report are those of the author and should not be construed as an official DoD position, policy or decision.

Contents

ACKNOWLEDGEMENTS

EXECUTIVE SUMMARY i
    The Range of Regional Nuclear Threats v
        Nascent or Tier One Capabilities: Limited Retaliatory Capability v
        Militarily Operational or Tier Two Capabilities: Multi-Salvo Capability vi
        Mature Tier Three Capabilities: Assured Retaliatory Capability vii
        The Shield/Sword Challenge vii
    US National Military Response Options viii
        The Status Quo, Option A: viii
        Moderate Adaptation Strategy, Option B ix
        Aggressive Adaptation Strategy, Option C x
    Overview x

I. THEMES FROM THE HISTORY OF NUCLEAR PROLIFERATION 1

II. THE STRATEGIC CHALLENGE OF EMERGING NUCLEAR POWERS 3

III. ALTERNATIVE NATIONAL SECURITY RESPONSES 7
    Option A 7
        Doctrine and Concepts of Operation 7
    Option B 8
        Doctrine and Concepts of Operation 8
    Option C 9
        Doctrine and Concepts of Operation 9

IV. PREPARING FOR NUCLEAR OPERATIONS: COLD WAR LESSONS LEARNED 11

V. RESPONDING TO A REGIONAL NUCLEAR CHALLENGE: THE STATUS QUO, OPTION A 15
        Enhanced Counterforce Investments 15
        All Weather Precision Guided Munitions 16
        Persistent Attack Munitions 16
        All Weather Precision and Persistent Surveillance and Targeting 16
        Enhanced Active Defense Investments 17
        The National Security Space (NSS) Architecture 19
        On EMP and High Altitude Nuclear Detonation (HAND) 20
        Enhanced R&D and Training 21
        Enhanced Expeditionary Capability 22
    An Overview 23

VI. "DEFEATING" A NUCLEAR-ARMED REGIONAL POWER: A MODERATE ADAPTATION STRATEGY, OPTION B 25
        Requirements for Option B 25
        The Dynamic Regional Nuclear Threat 25
        Counter-Nuclear Campaign Requirements 27
        Persistent Reconnaissance-Strike 28

      A New Generation of Earth Penetration Warheads.. 31
      Resurrecting Joint Counter-Nuclear Campaign Training............................... 32
      Active Defenses.. 32
      More Robust C4ISR... 34
      Preparing for EMP and HAND Attacks 36
      On Ground Force Operations.. 37
      Operating from Stand-off Distances.. 37
  Summation.. 38

VII. AN AGGRESSIVE ADAPTATION STRATEGY OF NEUTRALIZATION AND DEFEAT: OPTION C... 41
      On the Conduct of a Counter-Nuclear Campaign.. 41
      On Counterforce 41
      On Active Defenses 41
      Building a robust C4ISR infrastructure 41
      On EMP and HAND "Taxes"... 42
      Large Scale Follow-on Ground Operations 42
  A New Model Army and Marine Corps 43
      On Heavy, Medium, and Light Brigade Combat Teams (BCTs)..................... 43
      Refocusing the FCS Program 45
      On Vertical Maneuver Operations 45
      The Joint Heavy Lift (JHL) Concept 45
      Precision Air Drop 46
      The Semi-Buoyant Air Ship.. 47
      Expanding Seabasing 48
      Protecting Regional Allies.. 49

VIII. COUNTER-NUCLEAR CAMPAIGN OPTIONS: WHAT APPEARS FEASIBLE? 51
      Nascent or Tier One Capabilities: Limited Retaliatory Capability 51
      Militarily Operational or Tier Two Capabilities: Multi-Salvo Capability 51
      Tier Two Modernization and Response Options 52
      Managing Tier One and Tier Two Powers 52
      Constructing a Way Ahead 53
      Counter-Nuclear Campaign Requirements... 53
      New Model Combined Arms Operations.. 54
      Mature Tier Three Capabilities: Assured Retaliatory Capability................... 56

IX. CONCLUSION 59
      Summing Up.. 60

APPENDIX A: A BRIEF HISTORY OF NUCLEAR PROLIFERATION 63

APPENDIX B: HISTORY OF NUCLEAR OPERATIONS 69
  The Soviet Experience... 71

APPENDIX C: NUCLEAR WEAPONS EFFECTS 73
  Blast Effects 73
  Prompt Ionizing Radiation Effects... 73
  Thermal Radiation Effects... 73
  Local Radioactive Fallout Effects 75

Electromagnetic Pulse 75
   Introduction 75
   Physical Mechanics.. 76
   Effects of Hemp... 78
   Effects on Ground Systems 79
   Hemp Effects on Other Systems 80
   Preventative Measures.. 80
Using Scaling Laws to Estimate Nuclear Weapons Effects................................ 83

ACKNOWLEDGEMENTS

This paper was authored for the Center for Strategic and Budgetary Assessments (CSBA) by Mr. Peter Wilson and Mr. Elbridge A. Colby. Mr. Wilson is a senior professional staff member of the RAND Corporation, where he writes on a variety of national defense issues. Mr. Colby, formerly on the Commission on the Intelligence Capabilities of the United States Regarding Weapons of Mass Destruction, is an adjunct staff member at RAND. Their work was vetted and reviewed by CSBA's Vice President for Strategic Studies Bob Work, and Senior Fellow Barry Watts.

The analysis and findings presented here are solely the responsibility of CSBA.

EXECUTIVE SUMMARY

In order to sustain the current international system organized around American-led alliances, the United States may need to be able to confront challenges posed by revisionist powers armed with nuclear weapons. Immature or transitional nuclear powers are likely to pose especially pressing problems for US strategy and military planning over the coming decades. In light of this probability, the United States should develop the capability both to confront and, at least in a limited sense, defeat such powers while also preventing or deterring them from employing nuclear weapons for decisive effect. Such a balancing act will require a sophisticated set of capabilities and equally sophisticated planning, posturing, and action. This study will examine several different possible responses, each with a correlative set of capability requirements. The first option is to maintain the status quo with its brittle binary responses to nuclear threats: inaction or nuclear retaliation. The second option would invest in capabilities that allow the US to defeat an adversary willing to use its nascent nuclear arsenal. The last option is an extensive program intended to permit the US to conduct operations across the military spectrum in the face of significant nuclear use by an opponent.

The bottom line is quite simple: the United States should have military and technological options to deal with emerging nuclear powers. However, investing in meaningful response options would require the US political leadership to acknowledge that the current status quo strategy – with its focus on preparing for strictly conventional regional contingencies – is dangerously inadequate. This may be especially true in dealing with emerging nuclear states, such as the Islamic Republic of Iran, that have a strongly revisionist geo-strategic agenda. On the other hand, if the United States decides against making the investment to adapt to these emerging nuclear powers, it calls into question the central rationale for continuing a massive and sustained investment in high technology conventional capabilities since few will wish to fight the United States on its own terms. In a world of nuclear-armed adversaries, forces optimized to fight *only* conventionally-armed regional powers would seem to have little utility.

Trends Toward Nuclear Proliferation: Nuclear weapons capability has slowly spread through a process of punctuated equilibria.[1] The history of nuclear weapon proliferation has seen periods of rapid emergence of several nuclear-armed states followed by periods of relative quiescence. (See Appendix A for a greater explanation of this phenomenon). Despite pessimistic expectations, extensive proliferation did not occur during the Cold War. The United States and the Soviet Union prevented this outcome through the containment of regional contests within the larger superpower rivalry and their provision of security guarantees to likely developers of nuclear weapons. Regardless, several states elected to develop modest independent nuclear capabilities for reasons including prestige, independence, and deterrence. In the post-Cold War era, the fundamental dynamics of nuclear proliferation changed: regional competitions lost their restraints, superpower patronage waned, and US conventional capability became overwhelmingly dominant.

---

[1] For a description of this phenomenon see Roger C. Molander and Peter A. Wilson, " On Dealing with the Prospect of Nuclear Chaos," *The Washington Quarterly*, Summer 1994.

In this environment, obtaining a nuclear weapons capability became a more attractive proposition for a number of nations, particularly "rogue states" on the US regime "hit list," such as Iran and North Korea. For these regional powers at odds with the United States and bereft of a superpower protector—and unable to duplicate the scale of US conventional power—nuclear weapons promised to enable them to deter hostile US action and/or regional rivals through efficient, cost-effective, asymmetric means. Further, based on the past pattern of punctuated equilibria, proliferation to these nations seems likely to spur other regional powers to obtain nuclear weapons to deter, defend, and balance against them.[2]

In light of the stunning superiority of American non-nuclear arms, US preparedness to fight a war involving nuclear weapons has atrophied. Further, the starkest challenge to US military dominance in the post-Cold War period appeared not from the high end of the technological spectrum, but from insurgency and terrorism, leading to calls for an increase in manpower and funding for the Army and Marines instead of preparation for a nuclear, biological, or chemical (NBC) battlefield.

Trends suggest, however, that the United States will be challenged by new nuclear-armed powers seeking to defeat US conventional dominance by going "over" rather than (or in addition to) "under" it. The retired Chief of the Indian General Staff put it neatly several years after the Persian Gulf War of 1991: "Never fight the United States without nuclear weapons."[3] Nuclear

---

[2] The recent National Intelligence Estimate (NIE) on the status of the Iranian nuclear weapons program asserts with "high confidence" that Iran's weapons program has been "halted" since 2003. The authors inside the Intelligence Community (IC) believe that the Iranian leadership made a risk benefit calculation that the downsides, including the risk of economic and military sanctions by the United States and its allies, outweighed the benefits of an early weapon capability through a clandestine crash effort. On the other hand, the NIE acknowledges that the large scale "civilian" centrifuge enrichment program continues to mature, with the potential to produce sufficient fissile material for a nuclear weapon sometime between 2010 and 2015, a point emphasized by several distinguished commentators, including Henry Kissinger and Dennis Ross. Not surprisingly this IC "bombshell" has prompted a vigorous debate inside and outside of the Bush Administration as to whether the Iranian leadership is prepared to negotiate a "grand bargain" to make permanent the weapons program freeze and to lock down, via the IAEA, the ongoing fissile material production programs. Even with an internationally negotiated success, however, the US may have to accept that Iran will acquire through a robust "civilian" nuclear program a virtual nuclear arsenal, or break-out option. After all, the Iranian missile modernization programs that include the development of long-range solid propellant ballistic missiles continues to move apace. See National Intelligence Estimate, "Iran: Nuclear Intentions and Capabilites," Director of National Intelligence (DNI), November 2007. Also, see George Friedman, "The NIE Report: Solving a Geopolitical Problem with Iran," STRATFOR, December 3, 2007.

An ambigious news story has emerged in Northeast Asia, where the North Korea in 2007 agreed to shut down its nuclear weapons program. Currently, intense negotiations are underway to ascertain whether the North Korean regime will "come clean" and divulge the extent and location of its ongoing nuclear weapons program. As of the end of 2007, the DPRK has taken action to dismantle its small production reactor; on the other hand, it did not meet the deadline to reveal the full extent of its fissile material production programs and associated weapons program. It is noteworthy that in both the Iran and North Korean cases, financial, economic, and regional security incentives may prove sufficient to freeze – if not reverse – these programs. Libya's decision to abandon its nuclear program in return for diplomatic recognition and the opportunity to engage in global trade in energy can be seen as a model of a state ending its active nuclear weapons program.

[3] Cited in, Robert Manning, "The Nuclear Age: The Next Chapter," *Foreign Policy*, Winter 1997-1998. Former Chief of the Indian General Staff General K Sundarji wrote a book, *Blind Men of Hindoostan: Indo-Pak Nuclear War* (New Delhi: UBS Publishers' Distributors Ltd., 1993), that argued for an overt Indian nuclear capability and a national security council to manage the civilian and military command and control of this arsenal. Since the Indian nuclear tests and the overt emergence of the Indian and Pakistani nuclear capability, the Indian government has created a NSC. India's civilian leadership keeps very tight control over the nuclear arsenal. In turn, Pakistan has

weapons offer several benefits for countries seeking to take on or defy the United States. As the "absolute" weapon, their destructiveness provides an unmatched deterrent and/or terror capability. Such weapons, especially second generation weapons in moderate to large numbers with reasonably reliable delivery systems, offer effective deterrent capabilities. A North Korea or Iran stocked with survivable long-range ballistic and cruise missiles armed with nuclear warheads would likely be much less vulnerable to an American-led invasion or regime change. Furthermore, such weapons are relatively cheap, especially compared to the cost of achieving the same strategic objective of greater regional influence with highly trained, high technology conventional forces.

American Vulnerability: A significant disjuncture exists, therefore, between the increasing likelihood that US adversaries may possess nuclear weapons and American lack of preparedness to fight them. Many in US security policy circles dismiss the problem of fighting against a nuclear-armed opponent as overblown, though some accept the implications of proliferation and seem willing to accept a scaling back of American commitments abroad from their post-Cold War zenith. Others, however, argue that the US could continue its policies of assertive regional intervention and influence even in the face of proliferation. This group believes that a mixture of the taboo against nuclear weapons and the fear of US retaliation will prevent adversaries from crossing the nuclear threshold. The immediate demands of the large protracted counter-insurgencies in Iraq and Afghanistan contribute to the lack of focus on this problem, as does the need to prepare to fight conventional conflicts on the Korean Peninsula and elsewhere. Further, the inertial tendencies of the Pentagon planning process favor a continuation of the present program mix with its focus on high-technology conventional weapons.[4]

The situation, therefore, appears ripe for a strategic surprise in which a US opponent would employ nuclear weapons to stymie US and/or allied action. Several potential adversaries appear resolved to acquire a nuclear capability in order to deny, deter, or in key respects defeat the United States in a strategic conflict.

At the operational level, however, the US military appears unprepared to fight against a regional nuclear power. Furthermore, despite the fact that the planned US response to nuclear use by an opponent involves nuclear retaliation, there is strong evidence, uncovered in many RAND table top exercises during the early 1990s, of viscerally strong antipathy among policymakers against using nuclear weapons.[5] Given the military's apparent incapacity to fight a war against a nuclear

---

created a nuclear command and control system that is dominated by the military leadership with the civilian political leadership in a clearly subordinate role.

[4] See Defense Science Board (DSB), *Nuclear Weapons Effects Test, Evaluation, and Simulation* (Washington, DC: Undersecretary of Defense for Acquisition, Technology and Logistics, April 2005), for an extensive discussion of the fading significance of nuclear effects resilience (hardening) in the joint requirements process for current and next generation weapons and C4ISR systems.

[5] See Marc Dean Millot, Roger Molander and Peter A. Wilson, *"The Day After..." Study: Nuclear Proliferation in the Post-Cold War World*, Volumes I,II&III (Santa Monica, CA: RAND Corporation, MR-266-AF, 1993) for a very early post-Cold War analysis of the challenge of a nuclear-armed regional power. Since the initiation of that series, Roger Molander and Peter A. Wilson, along with a number of RAND colleagues, have developed a wide range of *"The Day After..."* exercises to explore the political and military challenges of further nuclear weapons proliferation.

power, combined with demonstrated policymaker reluctance to execute nuclear retaliatory threats, the US posture against prospective nuclear-armed powers is alarmingly brittle.

Fighting in a Nuclear Environment: Both the United States and the Soviet Union tried to field forces capable of waging war with, and in the face of, nuclear weapons. Their experiences suggest several hypotheses about military operations against nuclear-armed regional adversaries. While these hypotheses do not provide clear guidance for how to move forward, they do suggest enduring realities, limitations, and possibilities when dealing with nuclear warfighting:

- *There are severe and intractable difficulties in developing forces for nuclear warfare*

- *In a conflict involving nuclear powers, the line between conventional and nuclear and between tactical and strategic will not always be clear.*

- *The current IT-intensive RMA both eases and complicates the problems associated with operating in a nuclear battlespace.*

- *Alliance dynamics significantly complicate the preparation for and the conduct of nuclear operations.*

- *Institutional interests will play a powerful role and may not align with national strategic requirements.*

(See Appendix B for further discussion of the history of nuclear operations.)

On the US Nuclear Arsenal – Although not the focus of this analysis, the size and evolution of the US nuclear arsenal will obviously have some bearing on meeting the challenge of projecting power against nuclear-armed regional powers. As will be discussed below, even the most robust non-nuclear adaptive strategy will likely continue to rely on some measure of nuclear deterrence through the threat of assured nuclear retaliation. Hence, the size and capability of the US nuclear arsenal looms over any strategic interaction involving the United States or its allies. The nature of the arsenal, especially its employability and suitability for limited operations, will therefore affect the United States' ability to conduct operations against nuclear-armed regional powers.

Unlike during the Cold War, the size and configuration of the US nuclear arsenal may now be largely decoupled from the evolution of the two other major nuclear-armed nations, the Russian Federation and the People's Republic of China. Currently the United States and the Russian Federation are committed to reducing their operational strategic nuclear arsenals down to approximately 2000 weapons by 2012 via the Moscow Strategic Offensive Reduction Treaty (SORT). Furthermore, the ratified Strategic Arms Limitation Treaty (SALT-I) treaty with its extensive verification provisions expires in 2009. Given these developments, the new calculus for developing the arsenal might focus less exclusively on countering Russian and Chinese forces and more on how the arsenal will affect the behavior of other nuclear-armed or potential nuclear-armed states.

Under any circumstances, then, the next administration will have to address to what degree a new generation of nuclear arms control and reduction regimes should be a part of the international agenda. Already some former senior members of the US national security community are seriously suggesting that the US consider a strategy leading to the abolition of all operational nuclear arsenals.[6] It is beyond the scope of this analysis to evaluate the feasibility – much less the attractiveness – of these proposals. A central assumption of this study is that such disarmament efforts will not solve the strategic problem of emerging nuclear-armed powers. Therefore, the challenges and possible solutions identified herein will almost certainly be central to the US national security agenda of the next decade.[7]

THE RANGE OF REGIONAL NUCLEAR THREATS

In a proliferated environment, the United States will need to be able to engage a variety of nuclear-armed opponents.

Nascent or Tier One Capabilities: Limited Retaliatory Capability

Some nations may elect to obtain nuclear capabilities strictly for strategic deterrent purposes. Reflecting this policy, their arsenals would likely be tailored for counter-value functions in order to assure a retaliatory capability. The emerging Tier One first- generation nuclear arsenal will be small in number, numbering fewer than two dozen warheads, and consist of first generation fission weapons with yields of no more than 20 kilotons. The primary means of delivery will likely be either jet fighter aircraft or first-generation liquid propellant ballistic missiles. Since their arsenals would be solely defensive, unrefined, and terroristic in nature, US dealings with these countries are liable to be limited in nature. Regime change would be taken "off the table" in many, if not most, scenarios. While US forces and policymakers would have to be concerned with the prospect of escalation to the nuclear level, limited military operations themselves would be unlikely to be radically altered.

States that appear to have adopted minimum deterrent postures – in other words, enough to protect the regime from an outright military overthrow – include China (during its early operational capability) and South Africa. North Korea's current capability, such as it is, probably reflects a minimum deterrent posture. This may reflect necessity more than intent, however. Iran is also likely attempting to acquire this type of initial capability.

---

[6] See George P. Shultz, William J. Perry, Henry A. Kissinger and Sam Nunn, "A World Free of Nuclear Weapons," *The Wall Street Journal*, January 8, 2007 and their follow-on article in the *Wall Street Journal* in January 2008 as examples of the renewed call for nuclear "abolition." Also see Michael Krepon, "Ban the Bomb. Really," *The American Interest*, January/February 2008. For a contrary view, see the recent white paper proposed for the 2008 NATO summit in Bucharest, "Toward a Grand Strategy for an Uncertain World, Renewing the Transatlantic Partnership." This paper, authored by several former NATO and French senior military leaders, including General Klaus Naumann, former Chief of the German Defense Staff, and General John Shalikashvili, former Chairman of the Joint Chiefs of Staff, suggests that nuclear proliferation may be so severe in the Greater Middle East as to warrant the Atlantic Alliance maintaining preemptive nuclear options against nuclear-armed states.

[7] The authors of this monograph note that a serious attempt to move toward nuclear abolition is fraught with considerable uncertainty. The challenge of defining what is "zero" and sustaining a global restraint regime in a world without operational nuclear arsenals is considerable, especially if there is a renaissance in civilian nuclear power prompted by strong concerns about the consequences of global warming.

v

Militarily Operational or Tier Two Capabilities: Multi-Salvo Capability

A Tier Two nuclear power will deploy, at minimum, a multi-salvo capability. It will probably invest in a longer range regional/strategic strike capability primarily through a fleet of mobile and protected ballistic missiles. An array of tactical/operational weapons designed to support military operations will complement this strategic capability. The Soviet nuclear expeditionary force deployed to Cuba in 1962, sans its thermonuclear warheads, is a model for that mix of capability: approximately 60 nuclear-armed intermediate range (IRBMs) and medium range ballistic missiles (MRBMs) and approximately 130 "tactical" warheads that could be delivered by fighter bombers, short-range cruise missiles, and short range ballistic missiles (SRBMs).[8]

Tier Two nuclear states most likely will seek nuclear weapons not only for deterrent purposes but also for operational use in varying missions. Some states may deploy lower yield, earth penetrating, neutron (anti-personnel), or other weapons tailored for straightforward battlefield purposes. Such weapons could be used against the whole spectrum of classic military targets, including deployed forces in the field, staging grounds, depots, airbases, ports, other bases, transportation networks, supply lines, and so forth. Such weapons could also be used for indirect military purposes, such as area denial, terror and disorder inducers, and against economic and social targets. States may also develop weapons capable of anti-technology effects designed to destroy or degrade an opponent's command, control, communications, computers, intelligence, surveillance, and reconnaissance (C4ISR) capabilities, including electro-magnetic pulse (EMP) and anti-satellite (ASAT) functions.

Israel, India, and Pakistan all appear to possess arsenals of this type. Both India and Pakistan may attempt to acquire a Tier Three-like qualitative capability through the use of "boosting" techniques without the use of testing. More challenging for such nations will be the development of fission-fusion thermonuclear weapons.[9] Already, Israel is alleged to have a nuclear arsenal

---

[8] For an excellent comtemporary account of the Cuban Missile Crisis, see Norman Polmar and John D. Gresham, *DEFCON-2: Standing on the Brink of Nuclear War During the Cuban Missile Crisis* (Hoboken, NJ: John Wiley & Sons, 2006).

[9] See Chuck Hansen, *US Nuclear Weapons – The Secret History* (New York, NY: Orion Books, 1987), for a comprehensive unclassified disucussion of the US nuclear weapons program. See pages 18-103 for a detailed discussion of the development of fission, boosted-fission, and multi-stage fission-fusion (thermonuclear) weapons. Boosting involves the use of a "neutron-initiating gun" or external neutron sources to facilitate the more efficient "buring" of the fissile material. This involves access to tritium and deuterium gases and a high performance trigger mechanism. With boosting, a fission weapon might have a yield between a factor of two and ten times that of a standard 20-kiloton fission device. The more complex and demanding multi-stage thermonuclear device emerged with the mastery of the Teller-Ulam radiation implosion technique. See Ibid., page 22, for a diagram of a notional multi-stage thermonuclear weapon. Although the basic concept is in the public domain, the development of a reliable thermonuclear weapon is likely to require overt nuclear testing. On the other hand, with the extraordinary advances in computing power, future Tier Two states may be able to graduate to Tier Three status without an overt test. Israel, India, and Pakistan are plausible candidates in that regard.

that includes thermonuclear weapons.[10] Given the scale of its investment in a fissile material production infrastructure, it is likely that Iran aspires to a similar capability.[11]

Mature Tier Three Capabilities: Assured Retaliatory Capability

Some states may elect to join the United Kingdom, France, and China as mature nuclear powers, as defined by: the possession of survivable, second-generation thermonuclear warheads; sophisticated command and control (C2); and at least one method of transcontinental delivery.[12]

A typical Tier Three power will have diverse means to deliver its arsenal of 100-300 nuclear weapons. This will include second generation solid propellant ballistic missiles, both land and sea based, first generation low observable (LO) and supersonic cruise missiles, and long-range strike aircraft equipped with LO and supersonic cruise missiles. States with such a capability can derive a broader range of political and strategic options from their arsenals. Strategic interactions with such states are likely to tend towards caution and stability. Indeed, conflicts between the United States and mature nuclear powers will probably be highly stylized, much like interactions with the Soviet Union during the Cold War (especially after the Cuban Missile Crisis), with both sides showing considerable restraint in the use of military power. Major war between mature nuclear powers is highly improbable because no strategic aims are worth substantial nuclear retaliation. Strategic conflicts are thus likely to take on the character of "shadow war."[13]

The Shield/Sword Challenge

States such as Iran may develop nuclear weapons primarily as a deterrent "shield" but use their limited, retaliatory arsenal to cover aggressive "sword" operations through other means, including "irregular" or surrogate means of coercion.[14] Nascent nuclear powers like Iran may see

---

[10] See Anthony H. Cordesman, "Iran, Israel, and Nuclear War: An Illustrative Scenario Analysis," November 19, 2007, for an estimate of the current and future Israeli nuclear arsenal that likely includes thermonuclear weapons.

[11] The sheer scale of the Iranian nuclear production infrastructure suggests a national commitment to develop a capacity to produce sufficient fissile material to create a nuclear arsenal similar in size and character to India's or Pakistan's. For an analysis of the current status of the Iranian nuclear program see Anthony H. Cordesman and Khalid R. Al-Rodhan, *Iran's Weapons of Mass Destruction, The Real and Potentinal Threat* (Washington, DC: CSIS, 2007).

[12] The nuclear arsenals of the United States and the Russian Federation might be considered Tier Four capabilities with operational arsenals numbering in the thousands that can be delivered by a diversified force of transoceanic range missiles or aircraft. Tier Four arsenals are not included in this analysis.

[13] When the Cold War became a nuclear stand-off between the United States and the Soviet Union, much of their conflicts were conducted in the form of indirect, limited war and covert operations. This includes massive clandestine assistance efforts during the three major regional wars of the Cold War era – the Korean, Vietnam, and Afghan conflicts. See Xiaoming Zhang, *Red Wings Over The Yalu: China, the Soviet Union, and the Air War in Korea* (College Station, TX: Texas A&M Press, 2003), for a comphrensive account of the direct military assistance provided by the Soviet Union to Chinese and North Korean air forces during the Korean War. The full extent of this assistance was suppressed by the United States during the Cold War for fear of a US domestic call for a widened war with China. Similar clandestine operations occurred during the Vietnam and Afghan operations. Currently, the concept of state sponsored terrorism and counter-terrorism is consistent with the notion of indirect or "shadow" military operations.

[14] There is evidence to suggest that the Pakistani military thought that its emergent nuclear arsenal would shield it from large-scale Indian conventional escalation during the Kargil crisis of 1999, which was prompted by aggressive Pakistani military support for insurgents entering the contested border region of divided Kashmir. See Paraag Shukla, "Jammu and Kashmir (1947-Present)" in *Money in the Bank: Lessons Learned from Past*

openings for aggression because they believe their opponents will be sufficiently intimidated by the possibility of escalation to the strategic level to back down at lower levels of interaction.

Such a policy may be particularly troubling to the United States with respect to its far-flung and often ambiguous alliance commitments, which are more clearly justifiable in situations of nuclear monopoly than parity. Will nuclear possession by regional adversaries make the United States more vulnerable to "salami tactics" (whereby an adversary "slices" up an alliance through coercion) in areas not clearly vital to its strategic interests?[15]

US NATIONAL MILITARY RESPONSE OPTIONS

The Status Quo, Option A:

The current national security strategy, Option A, seeks to solve the problem of nuclear-armed adversaries through the existing set of capabilities, namely vastly superior conventional forces backed by the deterrent threat of nuclear or other retaliation in the event than an adversary uses weapons of mass destruction (WMD). While there are obvious advantages to the strategy, it is brittle. US forces are unprepared to respond operationally to nuclear use by an opponent other than with a massive and punitive conventional air campaign or through the use of our own nuclear arsenal. Wargaming evidence suggests an extreme aversion among US policymakers to the use of nuclear weapons and a strong desire to find non-nuclear responses.[16]

There is an element of bluff to the status quo posture and a future adversary may call that bluff. Given the increasing strength of the nuclear "taboo" among US Government and policy circles, combined with the notion of actual use seeming almost archaic after sixty years, such a call might leave the United States without realistic options.

Option A does provide, however, a mobilization base for the two more robust counter-nuclear options under consideration in this study. Specifically, substantial investments have been made to build robust counterforce and active defense capabilities. These are consistent with the current strategy of ensuring that the United States has overwhelming air supremacy to conduct joint combined arms warfare against regional adversaries conducting conventional military operations. The US has also devloped enhanced small scale (battalion-brigade echelon) joint forcible entry and expeditionary ground operations. As noted above, however, the basic rationale for these

---

Counterinsurgency (COIN) Operations, NDRI Occasional Paper, RAND OP-185-OSD, 2007, for a description of the Kargil crisis.

[15] For a discussion of the "salami tactics" problem, see Thomas Schelling, *Arms and Influence* (New Haven, CT: Yale University Press, 1966), p. 68.

[16] The results of RAND's over-fifteen-year *"The Day After..."* exercise series strongly suggest that US policy makers will be extremely reluctant to respond to a first nuclear weapon use, especially if it does not cause mass civilian casualties, with the prompt use of nuclear weapons against the major cities of the perpetrator. During the 1990s and through the turn of the century, most nuclear use exercises conducted by RAND, SAIC, CSBA, and others revealed a strong interest by both senior civilian and military leaders to find effective non-nuclear responses to first and limited nuclear use by a regional adversary.

modernization programs is to conduct short and violent Major Contingency Operations (MCOs) where nuclear weapons do not come into play.[17]

Moderate Adaptation Strategy, Option B

Option B takes as its premise that Option A is insufficient to provide either a credible deterrent or real warfighting capabilities against a nuclear-armed power in regional contests. The central operational assumption of this option is that, with proper and sustained investment in regional counterforce capabilities as well as in active and passive defenses, the United States can conduct a counter-nuclear campaign to neutralize the capability of an "immature" or Tier One opponent or at best an emerging Tier Two nuclear power.

Option B takes a "middle road" between the current strategy, Option A, which effectively ignores opponent nuclear capabilities, and a true attempt at full-fledged warfighting in a nuclear environment. Option B seeks to give greater plausibility to the US nuclear posture than is available under Option A – both for deterrent and warfighting purposes – while acknowledging that there are stark limits of nuclear use beyond which military operations are effectively impossible. It might be conceived of as a modern variant of Flexible Response, providing more tailored options to the US between an immediate resort to nuclear weapons in the event of a nuclear strike on the one hand and a more credible deterrent against such use by opponents on the other.

Therefore, Option B seeks to structure a US force able to conduct successful operations in the face of limited nuclear use without resorting to massive strategic nuclear retaliation, an option RAND research suggests is unlikely to be exercised by American policymakers.

Facing an enemy armed with a limited number of nuclear weapons that may successfully penetrate US defenses, the strategy posits that expeditionary forces in-theater must operate effectively in a dispersed, protected, flexible and mobile fashion from stand-off distances in order to minimize both their attractiveness as targets and damage in the event of a successful enemy strike. Nevertheless, under Option B the major fixed infrastructure targets of US regional allies and even of the United States itself, specifically cities, will still be vulnerable to nuclear attack. For this reason, deterrence likely will still play a vital role in controlling the level of conflict.

---

[17] US defense planning has made a strong de facto distinction between preparing for a nuclear battlefield and a battlefield "dirtied" with biological and chemical weapons. During the 1990s, the US made major investments in passive defense techniques to neutralize the military effects of the use of biological and chemical weapons. This investment was prompted in large measure by concerns about the prospect that North Korea had and would likely use chemical and biological weapons in a massive way during a second Korean War. See Albert Mauroni, "The Future of CBRN Defence." *Military Technology*, November 2007, for an excellent discussion about the current conceptual confusion between WMD and CBRN defense requirements. Mauroni writes, "Unless Russia or China starts another Cold War with the United States, nuclear weapons remain the only WMD threat. Any other nation using chemical or biological weapons cannot hope to develop, stockpile, and use the quantities of CB warfare agents against US forces necessary to create mass casualties (unless non-combatants are targeted), given modern counter-proliferation strategies and advances in protective equipment." Put simply, the CB warfare threat can be largely mastered through investments and training in passive defense measures unlike the more comprehensive threat of nuclear weapons.

Aggressive Adaptation Strategy, Option C

Option C proceeds from the belief that the United States will need to fight and win wars against nuclear powers, including in the face of significant nuclear use by the opponent. Option C's premise is that vital US interests cannot be adequately defended unless the United States can plausibly fight a nuclear war under most circumstances. A central assumption of this option is therefore that a sustained investment in counterforce capabilities as well as active and passive defenses will allow the United States to neutralize the nuclear capabilities of Tier One and Tier Two powers.

Like Option B, Option C acknowledges that a limited number of nuclear weapons may successfully get through and therefore that US expeditionary forces in-theater must operate in a dispersed/protected/flexible/mobile fashion from both stand-off distances and during deep offensive operations designed to defeat a nuclear armed regime. Even if the United States and allied military forces operate in a more protected and dispersed posture, however, the major fixed infrastructure targets of US regional allies, specifically cities, will still be vulnerable to nuclear attack. In this regard, the role of deterrence at the highest strategic level will remain critical.

OVERVIEW

The proper investment policy for nuclear warfighting should allow the United States to conduct military operations in the face of limited nuclear use. Such a policy would aim to prevent the US from being deterred or defeated by a power using relatively limited nuclear strikes. With the current approach, the US is vulnerable to a militarily disruptive attack or threat of attack by a power with a small number of nuclear weapons in a situation of asymmetric interests. A single or small number of nuclear strikes or credible threats of the same may strategically "freeze" the United States, caught between its highly vulnerable conventional forces and its almost unusable nuclear forces.

Developing the capability to frustrate such attacks will raise the threshold at which nuclear weapons buy strategic balance or parity with the United States, and will therefore preserve US strategic flexibility against Tier One and some rising Tier Two powers. Given the enormous potency of nuclear weapons this will, at the very least, require a marked change in the US military's approach to force planning, programming, and structure. The joint force will need to invest heavily in developing *highly* effective counterforce and active defense capabilities in order to deny opponents the ability to conduct successful nuclear strikes.

Assuming that "some will get through," US and allied forces will have to become more flexible, dispersed, mobile, and resilient, while employing other passive defense measures. US forces can realistically be equipped and structured to deal with limited nuclear strikes, especially in light of the IT-intensive RMA, which has already encouraged moves towards lightness, flexibility, mobility, and dispersion.[18] Indeed, structuring the military towards conducting limited nuclear

---

[18] A case can be made that there have been four revolutions in military affairs (RMAs) or ways of war during the 20th century. The first was the warfare innovations associated with the fighting vehicles that emerged from World War I and are a central feature of all "modern" military establishments. The second RMA was the emergence of nuclear weapons and their associated long-range means of delivery. The third RMA was the concept of Maoist "people's war" that emerged during the Chinese war with Japan and subsequent Chinese civil war. The fourth RMA was the IT-intensive RMA that has been misleadingly referred to as "the RMA." One hypothesis suggests that these RMAs, or ways of war have and will interact in a neo-Hegelian form of thesis and antithesis. This hypothesis

x

operations may have the corollary benefit of also preparing it to fight opponents with long-range and massed guided munitions and other tools of network centric warfare. Given that several potential rivals at the Tier Two and Tier Three level of nuclear capability are likely to pursue these as well as nuclear weapons, investments and reforms that address both emerging threats are certainly preferable.[19]

The objective of deterring an enemy from using limited nuclear strikes—and, if necessary, defeating one that has—should drive US military policy decisions. Moreover, such a program may also deter "on the fence" powers from even obtaining nuclear weapons. Conversely, such capabilities may incentivize Tier One nuclear powers to develop their arsenals to Tier Two or even Tier Three levels. Though this would be unfortunate, the consequences may not be all bad, as mature nuclear powers are likely to adopt superior command, control, security, and survivability practices. Further, US reticence about nuclear retaliation is likely to be much lower in the face of substantial nuclear use—the kind required to cross the threshold above which US military forces would be practically unable to operate.

The program advocated here shares some similarities with a modest version of the Flexible Response policy proposed to replace the Massive Retaliation doctrine of the 1950s. Like Flexible Response, this approach advises firming US capabilities at the lower levels of the nuclear escalation ladder in order to render more elastic the otherwise brittle commitment to massive nuclear retaliation in the face of limited nuclear use by an opponent. The policy proposed herein does not suggest that the United States should be obligated to meet enemy action symmetrically – nuclear retaliation, massive or otherwise, may be the appropriate response – but posits that the United States should have a broad spectrum of options to respond to nuclear use.

The United States will likely soon face a world containing new nuclear powers which have obtained these weapons in order to deter, deny, or defeat US efforts to secure American and allied interests. If current policy persists, the US may very well encounter limited nuclear war scenarios where the choice is between being deterred or defeated or employing nuclear weapons. The policy laid out here is an attempt to avoid this binary decision.

---

suggests that out of the interaction of the four RMAs of the 20th century, there may emerge a fifth RMA. This analysis does not suggest that either Options B or C is a fifth RMA; rather, it is an attempt to find an adaptation strategy to address the unfinished consequences of the emergence of nuclear weapons as a second RMA.

[19] Israel faced the worrisome prospect during the Second Lebanon War in July-August 2006 that both irregular and nation-state militaries using guided munitions and network-centric operations may gain a major military advantage. See Uri Bar-Joseph, "Israel's Military Intelligence Performance in the Second Lebanon War," *International Journal of Intelligence and Counterintelligence*, Vol. 20 No. 4, 2007.

I. THEMES FROM THE HISTORY OF NUCLEAR PROLIFERATION

Nuclear weapons capability has spread through a process of punctuated equilibria. Despite pessimistic expectations, extensive proliferation did not occur during the Cold War. The United States and Soviet Union retarded proliferation by containing regional contests within the larger superpower rivalry and by providing security guarantees to likely developers of nuclear weapons. Regardless, several states elected to develop modest independent nuclear capabilities for reasons including prestige, independence, and deterrence. In the post-Cold War era, the fundamental dynamics changed: regional competitions became more fragmented, superpower patronage waned, and US conventional capability became overwhelmingly dominant. In this environment, nuclear weapons became a more attractive proposition, particularly for "rogue states" such as Iran and North Korea – regional powers at odds with the United States and lacking a superpower protector. For these nations, nuclear weapons promised to enable them to deter hostile action by the United States or regional rivals through efficient and asymmetric means. Further, based on the past pattern of punctuated equilibria, proliferation by these nations seems likely to spur their neighbors to obtain nuclear weapons to deter, defend against, and/or balance them.

The Importance of the Nuclear Umbrella—Extending the Benefits of Nuclear Possession Without the Means: The single most important, though by no means sole, determinant in the history of proliferation has been the US and, to a lesser extent, Soviet security "umbrellas" or extended deterrent commitments.[20] These "umbrellas" retarded proliferation by extending the benefits of a nuclear capability to allied nations without disseminating the means. Thus the technology of nuclear arms proliferated as a strategic reality, but control remained in only a few hands. Naturally, the major issue for allied countries was whether the actual possessors of the weapons were reliable trustees, which explains the fervency of the NATO debates about the credibility of the American extended deterrent commitment in Europe. In the wake of the fall of the Soviet Union, however, a large number of states operate in an "unincorporated" nuclear environment, in which they are vulnerable to the US and other states across the conventional and nuclear spectrums.

Proliferation's Political Context—the Anxiously "Unincorporated" as First Movers: States that have elected to develop nuclear weapons are largely those that perceived themselves to be sufficiently threatened by, isolated from, or suspicious of the "manager/trustees" of the global security system to require the advantages that nuclear weapons are seen to confer.[21] States that

---

[20] For the history of why some US allies were dissuaded from obtaining nuclear weapons, including Japan, Germany, Taiwan, and South Korea, see Kurt M. Campbell et al, eds., *The Nuclear Tipping Point: Why States Reconsider Their Nuclear Choices* (Washington, DC: Brookings Institution Press, 2004).

[21] Most states seem to have obtained nuclear weapons primarily for what Scott Sagan calls "security" reasons. While domestic and prestige considerations have played into these decisions, especially in Britain, France, India, and Pakistan, security concerns seem to have been paramount in most countries, especially those about which the United States is most anxious. For further discussions of this topic see, Scott Sagan, "Why Do States Build Nuclear Weapons?: Three Models in Search of a Bomb," *International Security*, Winter 1996-1997, 54-86; Stephen Cimbala, *Nuclear Strategy in the Twenty-First Century* (Westport, CT: Praeger, 2000), esp. chapter 3; Kurt M.

are, in other words, anxiously "unincorporated" or "under-incorporated" into a credible security arrangement with the arbiters of the global system are likely to be first movers on developing nuclear weapons (China, North Korea, Iran, Israel, India, Pakistan). Conversely, vulnerable states that are incorporated into strong security arrangements have largely been dissuaded from obtaining an independent capability (NATO countries, Japan, Republic of Korea (ROK)).

Proliferation's Political Context—the Follow-On Effects of First Movers: As "first mover" states is likely to exacerbate the already difficult extended deterrent relationship between "trustee" nuclear states and allies threatened by the newly nuclear-capable state.[22] The French reaction to the Soviet development of nuclear weapons and their discomfort with American "trusteeship" was the first example of this sort of dynamic.[23] Along these lines, possession of nuclear weapons by North Korea and Iran will strain the nuclear umbrella relationship by intensifying the differences in security interests between the US and its allies in Northeast Asia and the Middle East. It is unclear whether regional allies will view American guarantees as sufficient for their security needs in the face of a regional rival in possession of nuclear weapons.

The Cascade Effect and Punctuated Disequilibrium: Nations that had previously accepted US nuclear trusteeship may individually elect to obtain a nuclear capability in the face of "rogue" state possession. Such states might include Japan, South Korea, Egypt, Saudi Arabia, and Turkey. Additionally, a classic arms race dynamic may emerge, in which a state's movement towards nuclear weapons capability prompts other states to do likewise as a region becomes nuclearized and American "trusteeship" becomes increasingly difficult to manage.[24] This increase may turn geometric once the ball gets rolling. Movements towards nuclear arms by either Japan or South Korea, for instance, would likely spur the other in the same direction, in turn prompting other nations in Asia to consider the option. Similar dynamics exist in the Middle East.[25] Given these unstable dynamics, the world may be on the verge of a sharp increase in the number of nuclear-armed powers.

Campbell, "Reconsidering a Nuclear Future: Why Countries Might Cross Over to the Other Side" in Kurt M. Campbell et al, eds., *The Nuclear Tipping Point: Why States Reconsider Their Nuclear Choices* (Washington, DC: Brookings Institution Press, 2004); Martin Van Creveld, *Nuclear Proliferation and the Future of Conflict* (New York, NY: The Free Press, 1993); and John Arquilla and Paul K. Davis, *Modeling Decisionmaking of Potential Proliferators as Part of Developing Counterproliferation Strategies* (Santa Monica, CA: RAND, 1993).

[22] For discussions of the likely criteria for obtaining nuclear weapons, see *The Nuclear Tipping Point: Why States Reconsider Their Nuclear Choices*.

[23] See, for example, Pierre Gallois, *The Balance of Terror: Strategy for the Nuclear Age*, R. Howard translation (Cambridge: Houghton Mifflin, 1961).

[24] Albert Wohlstetter posed this problem in his seminal article, "Nuclear Sharing: NATO and the N+1 Country," *Foreign Affairs*, Vol. 39 No. 3, April 1961. Keith Payne and Colin Gray have also made versions of this argument. See, Keith B. Payne, *Deterrence in the Second Nuclear Age* (Lexington, KY: University Press of Kentucky, 1996).

[25] For speculation about these possibilities, see Paul Bracken, *Fire in the East: The Rise of Asian Military Power and the Second Nuclear Age* (New York, NY: Harper Collins, 1999).

II. THE STRATEGIC CHALLENGE OF EMERGING NUCLEAR POWERS

In light of the declining threat of nuclear warfare, the IT-intensive RMA, and the protracted counterinsurgency operations in Iraq and Afghanistan, US military strategy and posturing has focused increasingly on conventional operations and low-intensity conflict. Grappling with a nuclear-armed enemy has taken a backseat to the extent that by the late 2000s the US military has become largely unprepared for military operations against a nuclear-armed opponent.[26] Yet, for a variety of reasons, adversaries of the United States and its allies will be strongly drawn toward asymmetric capabilities including nuclear weapons.[27] If the United States is to preserve the capability to act on behalf of its own interests and those of its allies in non-total war contexts, the US military will have to meet this challenge.[28]

From the central focus of military thinking in the 1950s, nuclear weapons became increasingly marginalized in real-world terms beginning in the 1960s with Flexible Response. This trend continued in the 1970s with the IT-intensive RMA, maturing in the 1980s under Reagan and Gorbachev, and coming to fruition in the post-Cold War era of US military dominance and Russian impotence. Purely conventional warfare, which had seemed dead to the strategists of the 1950s and fatal to early NATO planners, began to revive as American nuclear superiority decayed into parity, leaving a significant nuclear exchange appearing to many merely a recipe for mutual suicide. Western strategists began emphasizing the illogic of nuclear usage starting with the Flexible Response policy of the Kennedy-Johnson Administrations, which sought to give the Alliance the option to use limited means in defense of its interests.

The IT-intensive RMA strengthened this trend in the late 1970s as guided munitions, all weather sensors, and network-centric operations allowed NATO to defend Western Europe through conventional means alone for the first time. By the late 1980s, the advanced technology of the

---

[26] A clear signal of the US military's lack of focus on the nuclear operational challenge is the steadily diminishing – if not the complete waiving of – requirements for current and future weapons systems, sensors, and communications to be "hardened" to wide-area electromagnetic pulse (EMP) threats associated with atmospheric and high altitude nuclear detonations.

[27] See Peter A. Wilson, "Asymmetric Threats," in Hans Binnendijk, ed., *Strategic Assessment 1998: Engaging Power for Peace* (Washington, DC: National Defense University, 1998); and Stephen J. Cimbala, *Nuclear Weapons and Strategy: US Nuclear Policy for the Twenty-First Century* (London: Routledge, 2005).

[28] For discussions of this point, see the Post-Cold War Conflict Deterrence Study, *Post-Cold War Conflict Deterrence* (Washington, DC: National Academy Press, 1997), especially Appendices C, G, and H; Barry R. Schneider and Jim A. Davis, eds., *The War Next Time: Countering Rogue States and Terrorists Armed with Chemical and Biological Weapons* (Maxwell Air Force Base, AL: Air Force Counterproliferation Center, 2004); Barry R. Schneider, *Future War and Counterproliferation* (Westport, CT: Praeger, 1999); and Henry Sokolski, ed., *Fighting Proliferation: New Concerns for the Nineties* (Maxwell Air Force Base, AL: Air Force Counterproliferation Center, 1996).

Western powers seemed primed to defeat the Warsaw Pact without recourse to nuclear weapons, as evidenced by the decisive Israeli victory over Syria during the war in Lebanon.[29]

Operation Desert Storm demonstrated US convention master even more decisively at the beginning of the 1990s. With the collapse of the Soviet Union and the disappearance of any near-peer competitor to the United States, arguments for focusing on conventional superiority, supported by a military budget as large as the next dozen or more, carried more and more weight.

In light of this radical improvement in conventional capability and the stunning superiority of American non-nuclear arms, US preparedness to fight a war involving nuclear weapons atrophied.[30] Further, the most pressing challenge to US military dominance in the post-Cold War period came from low technology threats such as insurgency and terrorism, leading to calls for increased manpower and funding for the Army and Marines, as opposed to preparation for a high technology NBC battlefield. The immediate demands of the counterinsurgencies in Iraq and Afghanistan have contributed to the lack of focus on this problem, as did the need to prepare for conventional conflicts on the Korean Peninsula and elsewhere. Further, the inertial tendencies of the Pentagon planning process favored a continuation of the present program of conventional weapons.[31]

Trends suggest, however, that the United States will likely be challenged by new nuclear-armed powers seeking to defeat US conventional dominance by going "over" rather than (or in addition to) "under" it.[32] As the Chief of the Indian General Staff put it neatly several years after the Persian Gulf War of 1991, "Never fight the United States without nuclear weapons."[33]

Nuclear weapons offer several benefits for countries seeking to take on or defy the United States. As the "absolute" weapon, their unparalleled destructiveness provides unmatched deterrent and terror capabilities. In particular, second generation weapons in moderate to large numbers with reasonably reliable delivery systems (such as long-range ballistic and cruise missiles) would

---

[29] For a history of this development, see Lawrence Freedman, *The Evolution of Nuclear Strategy* (New York, NY: Palgrave MacMillan, 2003, 3rd Edition).

[30] See DSB, *Nuclear Weapons Effects Test, Evaluation, and Simulation*. For example, the EMP hardening requirement for the Army's premier modernization program, the Future Combat Systems (FCS) has been relaxed. The argument is that the extra cost of hardening this system of systems – an EMP "tax" of some 3-5% - is an unnecessary expense.

[31] Even with the revived interest in counterinsurgency and stability operations, the overwhelming focus of current procurement is in weapons systems designed to fight a high technology, conventional major regional contingency. The exception to this rule is the large countermine program, which includes the crash effort to deploy thousands of Mine Resistant Ambushed Protected (MRAP) armored vehicles. The MRAP is optimized to provide protection against IEDs and infantry small-arms including rocket propelled grenades (RPGs). It is not optimal in the face of an opponent armed with high performance, direct fire weapons such as anti-tank guided missiles (ATGMs).

[32] Michael Vickers and Robert Martinage refer to this effect as nuclear weapons' "strategic overhang." For discussion of this point and the vulnerability of conventional network-centric attack to nuclear use, see Michael G. Vickers and Robert C. Martinage, *The Revolution in War* (Washington, DC: Center for Strategic and Budgetary Assessments, 2004).

[33] Cited in Robert Manning, "The Nuclear Age: The Next Chapter," *Foreign Policy*, Winter 1997-1998.

make North Korea or Iran much less vulnerable to American-led invasion or regime change. Furthermore, such weapons are relatively cheap, especially compared to the cost of achieving the same ends with conventional means.[34]

As tools of politics, the mere threat of nuclear weapons can generate fissures in opposing alliances and within opposing nations. North Korea's bomb is already fracturing the American-led security community in Northeast Asia. Nuclear weapons could cow some nations into swearing off alliance with the US or, conversely, could force them closer into our embrace, potentially exacerbating tensions with fourth-party countries. Such weapons could also generate arms race dynamics that would destabilize finely balanced regions. Nuclear weapons also offer considerable bargaining leverage, evidenced by the preferential treatment accorded new entrants to the nuclear club. The intense emphasis put by the United States on preventing nuclear acquisition by "rogue" states – contrasted with the generous nuclear deal struck with former bad actor India – is indicative of the value of gaining the prize.

The impact on the military sphere is also considerable. Nuclear weapons offer the distinct possibility of negating the American advantage in conventional arms. Though significant elements in the US military establishment have convinced themselves nuclear weapons cannot be used for warfighting purposes, a sentiment expressed succinctly by then-Commander of Space Command General Charles Horner when he described nuclear weapons as "not useable," the truth is more complex and less comforting.[35] Though nuclear weapons certainly could be deployed as strategic terror weapons, they also have applications in more limited military contexts, including small scale tactical employment on the battlefield; attacks against US and allied staging grounds, depots, ports, and airfields; strikes focused against US technological superiority, such as EMP attacks designed to degrade and damage US C4ISR; demonstration detonations; counter invasion; and so forth. Though the United States managed to avoid any of these events during its standoff with the USSR, their fortuitous absence does not demonstrate their impossibility or even necessarily their implausibility.[36] Furthermore, the mere possession of nuclear weapons might have indirect dampening effects on US actions, enabling an opponent to use nuclear weapons as a shield and other forces as a sword.[37] Potential adversaries might, as the United States did on several occasions during the Cold War, consider the use of nuclear weapons in some limited fashion to be rational and cost-beneficial.[38]

---

[34] The counter invasion role of nuclear weapons should not be underestimated. Any attempt by the United States to conduct large-scale forcible entry operations against a nuclear-armed regional power is likely to be subject to nuclear attack. See Center for Strategic and Budgetary Assessments, (U) *Large-Scale Conventional WMD Elimination Operations Against a Regional State Adversary*, SECRET, Wargame report, April 2007.

[35] General Charles Horner, July 15, 1994 Press Briefing.

[36] See, on this point, Colin Gray, *The Second Nuclear Age* (Boulder, CO: Lynne Rienner, 1999).

[37] For some historical examples of this, see Richard K. Betts, *Nuclear Blackmail and Nuclear Balance* (Washington, DC: Brookings Institution Press, 1987).

[38] See National Security Decision Memorandum 242, available at www.fas.org/irp/offdocs/nsdm-nixon/NSDM_LIST.pdf, setting out policies for limited use of nuclear weapons. See also Presidential Directive 59 of 1981, available at http://www.jimmycarterlibrary.org/documents/pddirectives/pd59.pdf, doing the same.

A significant disjuncture exists, therefore, between the increasing likelihood that US adversaries will possess nuclear weapons and the American lack of preparedness to fight them. Many in US policy circles dismiss the problem of fighting against a nuclear-armed regional opponent as overblown though some accept the implications of proliferation and seem willing to accept a scaling back of American commitments abroad from their post-Cold War high water mark.[39] Others, however, argue that the United States could continue its policies of assertive regional intervention and influence even in the face of proliferation because nuclear weapons are, they contend, not useable.

The situation, therefore, appears ripe for a strategic surprise in which a US opponent would threaten or even use nuclear weapons to stymie US or allied action. On the opponents' side, several nations appear resolved to acquire a capability to deny, deter, or in key respects defeat the United States in a strategic conflict. At the operational level, the US military appears unprepared to conduct operations against a nuclear power, while at the US political level, despite the planned response to nuclear use by an opponent implicitly or explicitly involving nuclear retaliation, there is strong evidence, uncovered for instance in a RAND wargame in the early 1990s, of viscerally strong antipathy among policymakers against using nuclear weapons.[40] Given the military's apparent incapacity to fight a war against a nuclear power combined with demonstrated policymaker discomfort with executing on nuclear retaliatory threats, there exists a brittleness in the US posture against prospective nuclear-armed powers.[41]

---

[39] This has been the point of view of both John Mueller and Richard Betts. Others hope that the process of nuclear proliferation can be still be contained and will take heart from the prospect that both the Iranian and North Korean nuclear weapon programs can be frozen if not dismantled. See discussion of the recent Iran NIE above.

[40] See RAND's *"The Day After..."* and other policy research exercise experience.

[41] This brittleness has been emphatically emphasized by Keith Payne in *The Fallacies of Cold War Deterrence and a New Direction* (Lexington, KY: University Press of Kentucky, 2001).

III. ALTERNATIVE NATIONAL SECURITY RESPONSES

The following sections consider three alternative national security responses to the challenge of emerging nuclear-armed states. The first, or Option A, is the current posture which calls for only a minimal adaptation to this challenge and relies heavily on deterrence to neutralize the emerging nuclear arsenal. The second, or Option B, is a moderate adaptation strategy that places a very high reliance on current and next generation counter-force and active/passive defenses to deter, and if necessary, neutralize immature regional nuclear threats. The third response, or Option C, is an aggressive adaptation strategy that calls for a full spectrum of offensive and defensive capabilities to neutralize a regional nuclear power. The following is the strategic rationale for the three options in terms of how the Joint Operational Concepts (JOC) for Major Contingency Operations and Deterrence might be modified to meet this emerging challenge.

OPTION A

Doctrine and Concepts of Operation

The central assumption of Option A, the status quo, is that the Joint Operations Concept for MCOs and Deterrence[42] does not need not major revision in light of nuclear proliferation:

- *Planning for Nuclear Opponents*: The number of emerging nuclear powers can be contained through a vigorous combination of nonproliferation policy and action. Should such efforts fail, the US can deter emergent nuclear armed states from using their arsenal in a militarily significant way through the threat of regime and/or societal destroying non-nuclear and, if necessary, nuclear retaliation.

- *Continuing Need for Strategic Deterrence*: Defenses will not play a major part in dealing with nuclear-armed regional adversaries. High value civilian targets of regional allies may not be fully protected from nuclear attack, and their protection may continue to rely upon the threat of large scale retaliation including the use of nuclear weapons.

- *Current Counterforce and Defensive Capabilities Sufficient*: The current investment in a combined offensive and defensive counter-nuclear capability should, through deterrence and/or actual operation, be able to protect large scale expeditionary operations. Because of the reliance on deterrence and the emphasis on exclusion of nuclear weapons from conflicts, the United States may be better served under this option in not pursuing strong counterforce capabilities in order to avoid "use or lose" or "launch on warning" situations.

---

[42] See Joint Publication 3-12, *Doctrine for Joint Nuclear Operations*, Final Coordination (2), March 15, 2005, with its focus on the deterrence vice warfighting role of the US nuclear arsenal.

- *Necessarily Limited War Objectives*: Given the difficulties of a fully effective counter-nuclear neutralization campaign, the war aims of the United States in regional conflicts against nuclear-armed powers will almost certainly have to be limited. Future MCOs may take on features similar to the US experience during the Korean and Vietnamese wars where the United States did not use its full spectrum of military power to effect a strategically decisive outcome.

OPTION B

Doctrine and Concepts of Operation

Option B takes as its premise that Option A is insufficient to provide either a credible deterrent or real warfighting capabilities against a nuclear-armed power in regional contests. The central operational assumption of this option is that with proper and sustained investment in counterforce capabilities as well as in active and passive defenses, the United States can conduct a counter-nuclear campaign to neutralize a regional opponent armed with an "immature" or Tier One arsenal stocked with first generation fission weapons. Option B takes a "middle road" between the current strategy, Option A, which effectively ignores opponent nuclear capabilities, and a true attempt at full-fledged warfighting in a nuclear environment. Option B seeks to give greater plausibility to US nuclear warfighting than available under Option A – both for deterrent and warfighting purposes – while acknowledging that there are limits to nuclear use beyond which military operations are effectively impossible. It might be conceived of as a modern variant of Flexible Response, simultaneously providing more tailored options to the US in the event of a nuclear strike (as opposed to the immediate resort to nuclear retaliation) and a more credible deterrent against such use by opponents.

Option B therefore seeks to structure US forces to be credibly able to conduct successful operations in the face of limited nuclear use without recourse to massive, "strategic" nuclear response, an option RAND research suggests is unlikely to be exercised by American policymakers.[43] Facing an enemy armed with a limited number of nuclear weapons that may successfully penetrate US defenses, the strategy posits that expeditionary forces in-theater must operate effectively and decisively in a dispersed, protected, flexible, and mobile fashion from stand-off distances in order to minimize their attractiveness as targets and limit damage in the event of a successful enemy strike. Nevertheless, under Option B the major fixed infrastructure targets of US regional allies and even of the United States itself, specifically cities, will still be vulnerable to nuclear attack. For this reason, deterrence will still play a vital role in controlling the level of conflict.

Reflecting this strategic approach, the current JOC for MCOs will have to be revised in the following fashion:

- *Planning for Nuclear Opponents*: Acknowledge that future MCOs may involve opponents armed with and willing to use nuclear weapons.

---

[43] See RAND's *"The Day After..."* exercise results.

- *Counterforce and Active Defense a Paramount Capability*: A significant and effective counter-nuclear capability must be included in the capabilities suite of any large scale expeditionary operation. A major decision for Option B will be whether to execute counterforce strikes in advance of decisive operations, with the timing and strategic objectives of the operations dependent upon the outcome of the counterforce campaign. Since Option B charts a middle path, such a decision will need to be sensitive to the position and power of the adversary's nuclear arsenal.

- *Continuing Need for Strategic Deterrence*: Though defenses will play an increased role in Option B, high value civilian targets of regional allies may not be fully protected from nuclear attack, and their protection may continue to rely upon the threat of large scale retaliation including the use of nuclear weapons.

- *Setting War Objectives:* Under Option B, the United States will have to weigh whether an opponent's arsenal takes "victory" off the table as a possible option or whether the opponent can be militarily defeated at reasonable cost. In the former case, the United States will have to limit its war objectives along the lines of Option A.

OPTION C

Doctrine and Concepts of Operation

Option C proceeds from the assessment that the United States cannot adequately defend its interests unless it can fight and win wars against nuclear powers, including in the face of significant nuclear use by the opponent. A central assumption of this option is therefore that a sustained investment in counterforce capabilities as well as active and passive defenses will allow the United States to neutralize the nuclear capabilities of Tier One and Tier Two powers.

Like Option B, Option C acknowledges that a limited number of nuclear weapons may successfully get through and therefore US expeditionary forces in-theater must operate in a dispersed/protected/flexible/mobile fashion both from stand-off distances and during deep penetration operations. Even if the United States and allied military forces operate in a more protected and dispersed posture, however, the major fixed infrastructure targets of US regional allies, specifically cities, will still be vulnerable to nuclear attack. In this regard, the role of deterrence at the highest strategic level will remain critical.

Reflecting this strategic approach, the current MCO JOC will have to be revised in the following fashion:

- *Planning for Nuclear Opponents*: Acknowledge that most, if not all, future MCOs will involve opponents armed with a maturing nuclear arsenal.

- *Counter-neutralization*: A counter-nuclear campaign must precede or accompany any large scale expeditionary operation in order to achieve US objectives, up to and including victory.

- *A Theory of Victory*: Even with the limitations of a fully effective counter-nuclear campaign, the combatant commander, reflecting the strategic objectives of the USG, will not accept war outcomes that are closer to military stalemate than decisive military victory. Military victory will still be "on the table."

- *Extending Deterrence*: A central rationale of these revised JOCs on Major Contingency Operations and Deterrence is to signal to prospective nuclear armed opponents that the United States is prepared to use military force in a strategically decisive way even in the face of nuclear weapon use by that regional opponent. This would be intended to signal the continued strength and viability of American extended deterrence commitments to allies.

- *Continued Role of Deterrence*: Though defenses will play an increased role in Option C, high value civilian targets of regional allies may not be fully protected from nuclear attack. Their protection will continue to rely upon the threat of large scale retaliation including the use of nuclear weapons.

IV. Preparing for Nuclear Operations: Cold War Lessons Learned

The attempts of both the United States and the Soviet Union to field forces capable of waging war on a nuclear battlefield yield several lessons that should be borne in mind when thinking through the problem of how to conduct military operations against nuclear-armed regional adversaries. While these lessons do not provide clear guidance for how to move forward, they do suggest the enduring realities, limitations, and possibilities of nuclear warfighting. (See Appendix B for a more complete discussion.)

There are severe and intractable difficulties in developing forces for nuclear warfare. The United States eventually came to the conclusion that, beyond a certain point, military operations in a nuclear war would be extremely difficult, if not impossible, to conduct. The Soviets appear to have eventually reached a similar conclusion. The demands of survivability and effectiveness are conflicting and, on a nuclear battlefield, enormously intensified. The Pentomic Division is a paradigmatic example of these difficulties, and its failure came despite intense institutional interest in developing ways to conduct military operations in a nuclear war.[44] The vulnerability of US Navy carrier battle groups to saturation nuclear strikes during the Cold War is another such representative example.[45]

Conducting operations against an adversary with a large sophisticated arsenal is different than facing one with a small immature arsenal. A large scale nuclear war with a Soviet Union-type nuclear power would likely render military operations effectively impossible. A nuclear conflict with a country with a more limited arsenal, such as North Korea or an early Soviet Union or China, likely would allow significant space for military operations. Assessments that traditional warfighting would be possible in the 1940s and 1950s held greater plausibility than subsequently, when arsenals became much larger and more sophisticated.

In a conflict involving nuclear powers, the lines between conventional and nuclear and between tactical and strategic would not be clear. The "fog of war," incentive structures, differing interpretations of strategic categories, and other dynamics would all contribute to a great deal of vagueness and unpredictability regarding the use of nuclear weapons in a conflict. In an outright

---

[44] Fundamentally, the Pentomic Division concept was unworkable. It assumed that an infantry-type division could rely on a limited number of armored personnel carriers held at the division level to provide protected mobility for the maneuver battalions. The limitations of this design became evident during various field exercises that led to the emergence of the fully mechanized infantry divisions optimized for combat in Europe during the early 1960s. See John J. Midgely Jr., *Deadly Illusions: Army Policy for the Nuclear Battlefield* (Boulder, CO: Westview Press, 1986), for a history of the Army's attempt to adapt to the prospect of the sustained and mass use of nuclear weapons on a battlefield.

[45] Even in the Cold War, carrier battle groups might have survived repeated nuclear strikes if the Soviets' maritime surveillance system had been degraded and/or destroyed. On the other hand, the Soviets considered preemptive nuclear operations both feasible and very militarily desirable throughout the latter half of the Cold War.

nuclear exchange, these factors would be immensely intensified, but even in a declaredly conventional conflict there would be strong tendencies towards escalation.[46]

The current IT-intensive RMA both eases and complicates the problems associated with operating in a nuclear battlespace. A fundamental impact of the silicon-based technological revolution is the tremendous improvement in the effectiveness, reliability, and flexibility of conventional weapons and systems. Indeed, much of the appeal of the guided munitions/network-centric RMA lies in its promise to replace missions traditionally accorded to nuclear weapons with guided munitions and the like. This RMA enables lighter, less logistic-intensive, and more mobile conventional forces; precisely the attributes the planners of the 1950s identified as critical to operating in nuclear combat. That said, however, nuclear weapons in sufficient numbers are likely a "trump card" against advanced conventional capabilities, since they have orders of magnitude greater destructive power that can compensate for the sophistication of RMA systems. If available in large numbers, nuclear weapons can be used as brute force to defeat a technologically more sophisticated opponent. Furthermore, dominance in guided weapons/network-centric conventional warfare actually encourages future opponents to consider employing nuclear weapons to even the playing field. Finally, these high technology non-nuclear capabilities may be over-hyped, especially against targets that require ultra-reliable target location. The poor US performance against Iraqi SCUDs and mobile Serbian air defense system are two examples of the weaknesses exhibited by even the most modern conventional systems.[47]

Alliance dynamics significantly complicate preparing for and conducting nuclear operations. During the Cold War, US allies regularly and vehemently argued for markedly different approaches to nuclear operations. In Europe, especially after the 1950s, NATO allies, fearing the effects of nuclear combat on their territories, consistently pushed against the Alliance's focus on waging such a war. The effects of NATO's operation Carte Blanche in 1962, for instance, went a great distance towards convincing the West German government that nuclear war on its territory would be catastrophic, whoever ultimately prevailed. The emergence of large peace and anti-nuclear movements from the end of the 1960s also put immense political pressure on West European governments to shy away from nuclear warplanning.

Adding to the planning complexity within the NATO was the emergence of the UK and French independent nuclear forces. Over time, the British nuclear force became totally integrated into NATO/US nuclear operational planning. In turn, the British became increasingly dependent upon US assistance to sustain their national nuclear force of which the current Trident II Submarine

---

[46] See Barry Posen, *Inadvertent Escalation: Conventional War and Nuclear Risks* (Ithaca, NY: Cornell University Press, 1991).

[47] For further discussion of this point, see, Colin S. Gray, "Nuclear Weapons and the Revolution in Military Affairs" in T.V. Paul, Richard J Harknett and James J Wirtz, eds., *The Absolute Weapon Revisited: Nuclear Arms and the Emerging International Order* (Ann Arbor, MI: University of Michigan Press, 1998), pp. 99-134. See also James L. Geick, "Nuclear Weapons and the Revolution in Military Affairs," Masters Thesis at Naval Postgraduate School, 2000.

Launched Ballistic Missile (SLBM) program is the most obvious.[48] On the contrary, France developed a national nuclear concept of operation and remained independent of the NATO process. Although this raised considerable high-level political tensions during its emergence in the 1960s, NATO absorbed the French nuclear force in a political-strategic sense by the mid-1970s. The major challenges raised by the emergence of the French nuclear force inside the Atlantic Alliance suggest future challenges for the United States if major allies in the Greater Middle East or Northeast Asia conclude that they need an independent nuclear capability as well.

In Asia, anti-nuclear movements arose in South Korea in the 1980s and 1990s and pushed against the conduct of nuclear operations in a war on the Korean Peninsula. On the other hand, the decision to withdraw US nuclear forces in South Korea was prompted in large part by a US geo-strategic judgment that these "tactical" nuclear weapons were no longer necessary, especially after the Bush-Gorbachev September 1991 nuclear arms reduction agreement. The reverse dynamic, however, has also occurred, as when Fidel Castro encouraged Khrushchev and the Soviets to use nuclear weapons in the event of an American invasion of Cuba in 1962. Broadly, the experience of the last sixty years indicates that alliance dynamics add a significant complicating factor to planning for nuclear warfighting.

Institutional interests will play a powerful role and may not align with national strategic requirements. War on a nuclear battlefield may require capabilities and organizations radically different from conventional war. Yet such reforms must be enacted by established institutions that may find such changes inimical to their understanding of their core missions and characters. The history of the US Army during the 1950s is a classic example of how an entrenched institution can channel the demands of a new challenge, such as waging war on a nuclear battlefield, through its own preferred prisms.[49] Likewise, the US military's turn away from nuclear operations almost entirely in the 1990s may have reflected more the preferences of the United States and its armed forces than an objective analysis of the likely threats of the early 21st century. Military reform, always difficult, may be even more challenging in preparing for nuclear warfighting because such reform may threaten core values and programs of the combat services.[50]

---

[48] For a remarkable analysis of the relationship between the US and UK nuclear forces and the UK's proposed modernization program, see John Ainslie, *The Future of the British Bomb*, (London: WMD Awareness Program, Clydeside Press, 2006). The study provides a detailed analysis of the residual NATO forward deployed, dual-key "regional-strategic" air delivered nuclear capability.

[49] For excellent histories of this period, see Midgely Jr., *Deadly Illusions*; and Andrew Bacevich, *The Pentomic Era: The US Army Between Korea and Vietnam* (Washington, DC: National Defense University Press, 1986).

[50] The following from DSB, *Nuclear Weapons Effects Test, Evaluation, and Simulation*, p. 97, captures this phenomenon: "Nuclear survivability is important for the Combat Commander who expects to operate in a nuclear battlefield. However, most Combat Commanders apparently do not expect this to be the case, but they do believe that chemical, biological, and radiological threats will be faced. As a result most of the Initial Capabilities Documents, Analysis of Alternatives, Capability Development Documents, and Capability Production Documents do not address nuclear weapon environment survivability but do include chemical, biological, and radiological requirements. A recent survey of Air Force systems in the Information Retrieval Support System (IRSS) showed 16 systems had nuclear weapon hardness requirements in their System Requirements Documents, Mission Needs Statements, Statement of Needs, and Operational Requirements Documents, while 33 systems had chemical, biological, and radiological survivable requirements and the rest (approximately 100) were silent on the entire subject. Of the 16 systems, only 3 were not directly related to legacy strategic nuclear weapons systems. Combat

Commanders using non-strategic and non-nuclear systems (tactical) are more concerned with the day-to-day operational needs and likely battlefield situations that do not include the possibility of nuclear war. Likewise, many of the space assets are expected to operate in a non-nuclear weapons environment and only include the natural radiation environment requirements."

V. RESPONDING TO A REGIONAL NUCLEAR CHALLENGE: THE STATUS QUO, OPTION A

While the current US approach to dealing with nuclear-armed opponents has not yet resulted in any strategic defeat or embarrassment, such eventualities are increasingly likely as more powers obtain nuclear weapons and see opportunities to take advantage of binary American "all or nothing" commitments. Despite the brittleness of the status quo strategy, current and planned programs provide some of the counterforce and defensive capabilities required to confront a nuclear-armed adversary. In themselves, these capabilities are insufficient to confront a determined nuclear foe, however they do provide a foundation for the development of options A and B. The following section will discuss the status quo with an eye for how the US could develop these capabilities for more robust nuclear warfighting. There are major technological, operational, and organizational challenges to effect such a result.[51]

Enhanced Counterforce Investments

Developing a credible counterforce capability will require dramatic improvements. Historically, non-nuclear counterforce campaigns against mobile, hidden, and/or hardened targets have been unsuccessful. Noteworthy was the almost total failure of the "SCUD hunt" campaign during the first Persian Gulf War in 1991. Even with the extensive deployment of Special Forces units in the suspected Iraqi SRBM operating areas of western Iraq and the employment of more than 3500 tactical air sorties, the campaign recorded failed to knock out a significant number of Iraqi SCUDs – and may not have knocked out any.[52] To defeat a regional opponent armed with a nuclear arsenal, even of modest size, the performance of any counterforce campaign will have to improve radically if Option B is to be attractive for policymakers. The basic requirements for a successful counterforce capability are near-real time surveillance, tracking, and target acquisition; platforms that can respond very quickly to acquisition, acquire the target, and launch the appropriate weapons package against it; weaponry that can defeat, destroy, or disable enemy nuclear capabilities that are hidden, mobile, and/or hardened; and capabilities that can correctly assess whether a strike has been successful and take required follow-on actions.[53]

Many of the components of this improvement in capability already largely exist, as the US has made major progress since the first Persian Gulf War of 1991 through a massive investment in capabilities relevant to any future counter-nuclear campaign. The most important improvements include the following:

---

[51] See Barry R. Schneider, "Seeking a Port in the WMD Storm: Counterproliferation Progress, Shortfalls and the Way Ahead," *Counterproliferation Paper 29* (Maxwell Air Force Base, AL: Air Force Counterproliferation Center, May 2005).

[52] See Thomas A. Keaney and Eliot A. Cohen, *Gulf War Air Power Survey Summary Report* (Washington, DC: Government Printing Office, 1993), p. 91.

[53] See Barry R. Schneider, "Counterforce Targeting Capabilities and Challenges," *Counterproliferation Paper 22* (Maxwell Air Force Base, AL: Air Force Counterproliferation Center, August 2004).

All Weather Precision Guided Munitions

The emergence of the Joint Direct Attack Munitions (JDAM) represents the culmination of the precision aerial delivery "revolution" that has been underway for more than thirty years. Currently, the use of the navigation signals from the Global Positioning System (GPS) satellite constellation provide both long range artillery and rocket projectiles with a similar precision guidance capacity. Both the Excalibur GPS-guided 155-mm artillery shell and a guided version of the Multiple Launch Rocket System (GMLRS) have been used operationally in Iraq. Also, the Army has used the longer range Army Tactical Missile System (ATACMS) with unitary warheads and GPS guidance in Iraq.

Level of Performance: With JDAMs, the Air Force and Navy combat aircraft can strike at the full array of fixed and known targets at operational and strategic depth in all weather conditions, day or night. Further they provide the US Army and Marine Corps with non-organic close support fires.[54]

Persistent Attack Munitions

The third generation Tomahawk long-range cruise missile, the Tomahawk IV, provides a capability to loiter over a target area after launching. It provides a first generation capability to conduct unmanned persistent reconnaissance and attack.[55]

Level of Performance: The Tomahawk IV has limited endurance to conduct reconnaissance-strike operations at its full range of approximately 1500 miles. Further the current on board sensor is optimized to provide updates against fixed targets and not search for mobile, much less camouflaged, targets.[56]

All Weather Precision and Persistent Surveillance and Targeting

The development of the E-8C Joint Surveillance, Target Attack Radar System (JSTARS) provides theater and tactical commanders a long enduring aerial radar-based targeting platform.[57] This has been powerfully supplemented by the emergence of the MQ-1 Predator, RQ-4 Global Hawk, and MQ-9 Reaper unmanned aerial systems (UAS). They can carry a mix of electro-optical (EO) and synthetic aperture radar (SAR) sensors to provide all-weather reconnaissance.[58]

---

[54] For further information on the Joint Direct Attack Munition (JDAM), see http://www.globalsecurity.org/military/systems/munitions/jdam.htm.

[55] For a discussion of the evolution of the Tomahawk program toward the current Tomahawk IV design see Government Accountability Office (GAO), *Long-Term Implications of Current Defense Plans: Summary Update for Fiscal Year 2008* (Washington, DC: GAO/NSIAD-95-116, December 2007), especially Appendix I, "Tomahawk Improvements Address Limitations."

[56] For further information on the Tomahawk Land-Attack Cruise Missile (TLAM), see http://www.globalsecurity.org/military/systems/munitions/bgm-109.htm.

[57] For information about the Joint Surveillance Target Attack Radar System (JSTARS), see http://www.globalsecurity.org/intell/systems/jstars.htm.

[58] National Academy of Sciences, *Autonomous Vehicles in Support of Naval Operations* (Washington, DC: The National Academies Press, 2006).

Level of Performance: In combat environments without the threat of meaningful air defenses, these manned and unmanned surveillance aircraft can provide all-weather surveillance to the joint force. In the face of moderate to high air defense threats, these vehicles cannot operate over potential targets of interest and instead must operate in a stand-off mode. In the later case, the effectiveness of the aerial surveillance platforms decreases due to terrain masking of potential targets and the lower resolution of their radar and electro-optical sensors operating at longer distances.[59]

The combined effect of these broad advances in reconnaissance-strike capability is that the United States can readily attack and destroy all known targets on the surface of the earth, land or sea. With its undisputed air supremacy following a Suppression of Enemy Air Defense (SEAD) campaign, the US military can readily defeat conventionally equipped and organized armed forces under all weather conditions.[60] What these major advances in combat capabilities have not provided is military dominance against insurgent forces or nation states armed with hidden, mobile, or buried long-range ballistic and cruise missiles.

Enhanced Active Defense Investments

Highly effective active defenses will play a critical part in any successful variant of Option B. The United States will need to reduce its current vulnerability to ballistic and cruise missile attack if US forces are to be able to operate effectively in an NBC environment.[61]

The US military has made some significant strides forward in active defenses. After a very large and sustained investment in ballistic missile defense (BMD), the United States is now deploying operational systems with capabilities superior to the marginal capability of the Patriot-2 (PAC-2) surface-to-air missile system. These are the Patriot-3 (PAC-3), with a hit to kill interceptor, and the Navy's Standard Missile 3 (SM-3) with a hit to kill interceptor. The former was successfully used during the opening days of Operation Iraqi Freedom (OIF) to intercept a small number of *Al Samoud* SRBMs. The latter should have an initial operating capability (IOC) by the end of 2008. The Ground Based Interceptor (GBI) is a national defense system that became operational in Alaska during 2007. Similar to the SM-3, this is an exo-atmospheric hit to kill system. At present, there are three major programs to provide additional ballistic missile defense capability:

The Theater High Altitude Air Defense (THAAD) System – This long delayed Army program may have an IOC by the end of the decade. It will provide an upper layer to the terminal defense

---

[59] During the various conflicts with Serbia over the fate of Bosnia and Kosovo, the E-8C JSTARS had difficulty locating targets in mountainous terrain while standing off to avoid the threat of Serbian medium altitude air defenses.

[60] One of the more dramatic examples of the improved Air Force all-weather close support capability was the use of B-1Bs carrying JDAMS to attack maneuvering Iraqi mechanized forces at the height of a massive sand storm during Operation Iraq Freedom.

[61] See, for instance, Jeffrey A. Larsen and Kerry M. Kartchner, "Emerging Missile Challenges and Improving Active Defenses," *Counterproliferation Paper 25* (Maxwell Air Force Base, AL: Air Force Counterproliferation Center, August 2004).

capability of the PAC-3 with a hit to kill interceptor optimized to operate outside of the atmosphere.[62]

Airborne Laser (ABL) – This is another long delayed Air Force program that may provide a limited boost-phase intercept capability by the middle of next decade. This program faces major technological challenges and may not lead to an operational system.[63]

Kinetic Boost-Phase Interceptors – Although a land-based boost-phase interceptor program is no longer being funded, there is the promising Network Centric Airborne Defense Element (NCADE) program. This program is designed to develop modified versions of either the AIM-9 Sidewinder or the AIM-120 AMRAAM to allow for fighter aircraft to intercept ballistic missiles during their boost-phase. Further, there are plans to test an air launched version of the much heavier PAC-3 terminal interceptor.[64]

Joint US-Japanese BMD – The Government of Japan has signed on to a robust joint effort to develop and deploy a robust follow-on to the SM-3. Currently, this improved hit to kill system will be sea-based, but the Japanese may chose to deploy a land-based variant as well.[65]

Aside from these initiatives, there is an effort to improve our air defense capability, specifically against cruise missiles. Current and next generation fighter aircraft are and will be equipped with Active Electronically Scanned Array (AESA) radars with a greatly improved capacity to detect small targets such as next generation LO cruise missiles. The United States continues to modernize its arsenal of air-to-air missiles with next generation versions of the AIM-9 Sidewinder and the AIM-120 AMRAAM, the AIM-9X and the AIM120D, respectively. Both will be more effective against cruise missile class targets. There is the Army's Joint Land-Attack Cruise Missile Defense Network System (JLENS), a radar system on a tethered aerostat. This system provides wide area surveillance against low-flying aircraft and cruise missiles similar to the E-6 Airborne Warning and Control System (AWACS) and the E-2 Hawkeye radar

---

[62] See Nathan Hodge, "LM Wins Order for Production of THAAD System," *Jane's Defence Weekly*, January 10, 2007.

[63] For an overview of US missile defense programs and capabilities, see Steven A. Hildreth, *Missile Defense: The Current Debate* (Washington, DC: Congressional Research Service, 2005). See also Steven M. Kosiak, *Arming the Heavans: A Preliminary Assessment of the Potential Cost and Cost-Effectiveness of Space-Based Weapons* (Washington, DC: Center for Strategic and Budgetary Assessments, 2007).

[64] The first success of the NCADE program in December with a modified AIM-9 opens up the possibility that the large tactical fighter fleet could be employed in defensive combat air patrols (CAP) to either pin down and/or intercept mobile theater ballistic missiles. Noteworthy is the prospect that this new capability will reinvigorate the F-22's mission of air supremacy over a hostile airspace. See further discussion in Option C. Also see Stephen Trimble, "New Options Enter Race for Boost-Phase Intercept Weapon," *Jane's Defence Weekly*, May 31, 2006.

[65] For background on this effort, see Richard P. Cronin, *Japan-US Cooperation on Ballistic Missile Defense: Issues and Prospects* (Washington, DC: Congressional Research Service, 2002). Also, see Joris Janssen Lok, "Aegis Road Map US Navy, Lockheed Martin Line up Aegis Upgrades to Counter Increasingly Challenging Ballistic Missile Threats," *Aviation Week & Space Technology*, December 3, 2007. The joint US-Japanese BMD program will lead to a Block II variant of the SM-3 with an upgraded second stage motor and a next generation, higher performance hit-to-kill interceptor. If Aegis crusiers can operate off North Korea's coastline, then the Block II may provide a boost-phase intercept capability, if only against liquid propellent MR/IRBMs.

surveillance aircraft. Both are undergoing major upgrades to provide for an enhanced low altitude air defense capability.[66]

The National Security Space (NSS) Architecture

The central backbone supporting national and theater C4ISR systems is the current and planned national security space (NSS) architecture. This is the array of satellites that provide precision navigation and timing (PNT), earth observation, and global telecommunications. Currently the vital PNT mission is proved by the NAVSTAR GPS space constellation that operations at Medium Earth Orbit (MEO). Most earth observation, (i.e., space reconnaissance) satellites operate at Low Earth Orbit (LEO). Others such as some metrological and early warning satellites operate at Geosynchronous Earth Orbit (GEO). Most communications satellites operate at GEO but there is a family of satellites that provide mobile communications at LEO.[67]

This space services infrastructure is sustained by a NSS launch capability. Presently, this is provided by the United Launch Alliance (ULA) that operates two families of the Evolved Expendable Launch Vehicle (EELV), the Delta IV and Atlas V. This launch infrastructure operates on a "launch on schedule" concept of operations that allows for the steady modernization and replenishment of the three NSS space services described above.[68]

This architecture is very vulnerable to attack, especially those systems operating in LEO. The recent Chinese non-nuclear hit to kill anti-satellite (ASAT) test has highlighted this threat to satellites operating at LEO. The architecture will become even more vulnerable if a regional power uses nuclear weapons as either an ASAT or to generate an adverse operational environment.

At present, these emerging threats have prompted a debate within the space services and launch communities as to whether a new NSS architecture is needed. Currently, the Air Force is exploring the concept of Operationally Responsive Space (ORS). Two different approaches have been considered. The first is to develop a new generation of much smaller and less costly satellites that will be "launched on demand" by a new family of responsive and low cost space launch vehicles. The second is to consider the development of a new generation of space reconnaissance satellites that will operate in much larger constellations than the small number of very heavy and costly reconnaissance systems in use today. The first concept is focused on providing the national and theater users with a capacity to reconstitute some of the capability degraded or destroyed during a regional conflict. The second focuses less on reconstitution

---

[66] See Martin Streetly, "US Navy Rolls Out First E-2D Advanced Hawkeye," *International Defense Review*, June 1, 2007; and Martin Streetly, "Advanced Hawkeye Promises Quantumn Leap in US Navy's AEW capability," *International Defense Review*, August 1, 2006, for description of Navy AEW upgrades. See Martin Streetly, "Next-Generation AWACS Begin Mission System Flight Testing," *International Defense Review*, June 1, 2007, for a description of the E-3G variant of the E-3 *Sentry*.

[67] For a description of the National Security Space (NSS) satellite services see Forrest McCartney, *National Security Space Launch Report*, RAND National Defense Research Institute, MG-503-OSD, 2006.

[68] For a description of the evolution and current status of the EELV programs see McCartney, *National Security Space Launch Report*.

during wartime and more on the provision of more service in the field of interest of a theater command similar to the continuous coverage provide by the GPS navigation system.[69]

On EMP and High Altitude Nuclear Detonation (HAND)

When a nuclear detonation takes place at or near ground level a direct electromagnetic pulse from the explosion will damage components attached to large antennas and power lines within a limited range of the explosion, depending on yield. For example, the EMP effects for a 10-20 kiloton ground burst may only extend a few miles from the site of the detonation. A ground or low-altitude burst can also destroy computers within the range of the blast and thermal effects. It is also likely to affect communications networks, but the range of the disruption is very limited.

As described in Appendix C, the high-altitude EMP (HEMP) effects are less well understood, but could be substantial depending on height of burst and weapon yield. In considering HEMP effects, it must be kept in mind that assessments in this realm of nuclear weapons effects are derivative of a very limited number of high-altitude tests and that there is considerable debate and uncertainty about the magnitude of such effects on contemporary electronic systems.[70]

The emerging long-range strike, active defense and C4ISR capabilities described above are being developed with only modest emphasis given to the prospect that they will have to operate in a nuclear battlespace.[71] They may have to operate in an environment where one or more nuclear weapons have been detonated to generate wide area electromagnetic effects or detonated in space to "pump" the van Allen belt, which could degrade all LEO satellites.[72]

---

[69] For an evaluation of the ORS concept see McCartney, *National Security Space Launch Report*, pp. 39-43. Pedro Rustan of the National Reconnaissance Office (NRO) is one of the foremost proponents of a new generation of smaller and more numerous LEO reconnaissance satellites. Battlefield commanders will be provided with much more frequent coverage than with a large constellation of LEO surveillance satellites. This is the concept of the Space Based Radar (SBR) constellation. At present, the later program has been all but killed by Congress over technology risks, costs, and operational issues.

[70] The RAND Corporation did an assessment of the EMP threat in conjunction with the 2004 Congressionally mandated *Commission to Assess the Threat to the United States from Electromagnetic Pulse (EMP) Attack*. It found that for low-yield first generation fission weapons, the HEMP effect would be more disruptive in nature, for instance to system control and data (SCADA) system of the civilian power grid. Modern commerical electronics for automobiles could also be vulnerable because of their dense and low power solid state design, but they are likely to be resilient to low-yield HEMP effects because of their protected configuration against the electromagnetic interference (EMI) found in their normal operating environment. Other experts testifying before the commission and Congress expressed considerable alarm about the vulnerability of the US civilian IT infrastructure. Some have noted that while weapons systems and the narrow-band communications systems associated with Cold War era strategic nuclear forces are relatively well protected, US tactical weapon systems have emerged as vulnerable to nuclear generated electromagnetic effects.

[71] An important exception to this rule has been the continued investment in next generation EHF communication satellites, the Advanced EHF (AEHF), as a higher bandwidth follow-on to the MILSTAR constellation, which was designed to function in a nuclear disturbed environment. EHF provides a means to communicate through the ionosphere even if it is energized by high altitude nuclear detonations (scintillation).

[72] For more on the EMP issue, see *Report of the Commission to Assess the Threat to the United States from Electromagnetic Pulse (EMP) Attack*, (Washington, DC: Commission to Assess the Threat to the United States from Electromagnetic Pulse Attack, 2004). The Defense Threat Reduction Agency (DTRA) has studied the phenomenon of High Altitude Nuclear Detonations and has concluded that a 20 kiloton detonation at the appropriate altitude will energize the lowest van Allen belt and will radically reduce the lifetime of all unhardened satellites operating in LEO. See Samuel Glasstone and Philip J. Dolan, *The Effects of Nuclear Weapons* (Washington, DC: United States

A central assumption of current national security planning in Status quo, Option A, is that the potential for the use of nuclear weapons during a regional conflict is very low to non-existent. The presumption is that the US nuclear retaliatory capability will deter any regional power from conducting nuclear operations. From this assumption, there has been a rather consistent policy that the current and next generation weapons and C4ISR systems should not pay a significant "EMP tax" to provide for protection against this phenomenon.[73]

Enhanced R&D and Training

The US national security leadership has acknowledged the challenge of a counter-nuclear campaign since the first Persian Gulf War. Throughout the 1990s, the Defense Threat Reduction Agency (DTRA) became the lead agency to develop improved counter-force capabilities. The financial support for a variety of Advanced Concept and Technology Demonstrators was particularly noteworthy. These included the demonstration of a hard target smart fuse, an advanced earth penetrating warhead (EPW), and improved targeting concepts against mobile and hidden targets.[74] Further, DTRA has developed planning tools for the conduct of counter-nuclear campaigns including the Integrated Munitions Effects Assessment (IMEA), the Munitions Effects Assessment (MEA), and the Hazard Prediction and Assessment Capability (HPAC).[75]

With the reorganization of the combatant commands during the turn of the century, US Strategic Command (USSTRATCOM) took on the counter-nuclear mission as part of the broader counter-WMD campaign. To that end, the command developed jointly with DTRA the Counterproliferation Analysis Planning System (CAPS), a computer-based program at Lawrence Livermore National Laboratory where intelligence is collected, analyzed, and displayed on the WMD programs of states of concern. CAPS has become a key enabler of USSTRATCOM's capability to support regional combatant commanders in the event they require the development and execution of a regional counter-nuclear campaign.[76]

Departments of Defense and Energy, 1977, 3rd Edition), Chapter X, "Radio and Radar Effects," for a summary of the diverse nuclear effects on key elements of the IT-intensive elements of contemporary warfare.

[73] For discussion of this phenomenon see DSB, *Nuclear Weapons Effects Test, Evaluation, and Simulation*. Also see Ian Steer and Melane Bright, "High-Altitude Nuclear Explosions: Blind, Deaf, and Dumb," *Jane's Defence Weekly*, October 23, 2002.

[74] Noteworthy DTRA efforts include the "DIVINE" series of tests as part of the Hard and Deeply Buried Target Defeat (HDBTD) program. Emerging from this effort has been the 30,000 pound Massive Ordnance Penetrator (MOP). Currently, this non-nuclear Earth Penetrating Weapon (EPW) will be carried internally by the B-2A and externally by the B-52H with a IOC during 2008. See Bill Sweetman, "A MOP For Those Tough Clean-Up Jobs," *Ares, A Defense Technology Blog*, December 27, 2007, available at http://www.aviationweek.com/aw/blogs/defense/index.jsp?plckController=Blog&plckScript=blogScript&plckElementId=blogDest&plckBlogPage=BlogViewPost&plckPostId=Blog%3a27ec4a53-dcc8-42d0-bd3a-01329aef79a7Post%3a29e32390-8cc0-46d4-9fc5-41f041b2ae80.

[75] Schneider and Davis, *The War Next Time: Countering Rogue States and Terrorists Armed with Chemical and Biological Weapons*.

[76] Ibid.

Until the turn of the century, the counter-WMD challenge was given sufficient emphasis that the ROLLING SANDS exercise, initiated in the mid-1990s, became the premier joint training effort. This exercise was specifically designed to train the joint force in combined arms counter-WMD operations with the orchestrated use of counterforce and active defense assets through ever improving C4ISR capabilities. By 2003, these annual exercises ceased due to the personnel and financial resources being directed toward the demands of Operation Enduring Freedom (OEF) and Operation Iraqi Freedom (OIF). Further, high level national security focus shifted to the challenge of defeating the threat of improvised explosive devices (IEDs). This led to the multi-billion dollar Joint Improvised Explosive Device Defeat Organization (JIEDDO). Even the syllabus of the Army's National Training Centers (NTCs) shifted away from conventional combined arms training, tactics, and procedures (TTPs) that of dealing with the complex challenge of counterinsurgency and stabilization operations.[77]

Enhanced Expeditionary Capability

The Future Combat Systems (FCS) – The Army's premier modernization plan includes the re-equipment of roughly one third of the active Army (fifteen Brigade Combat Teams (BCTs)) with a new generation of C4, a wide range of air and ground unmanned vehicles, and a new family of armored fighting vehicles. At the beginning of this program, the Army hoped that the weight of the family of combat vehicles could be less than 20 tons. This would facilitate the use of selected FCS-equipped BCTs as mounted vertical maneuver (MVM) or "air mechanized" units. The BCTs would maneuver through the air via very large vertical takeoff and landing (VTOL) assault cargo aircraft to support distant (up to five hundred miles) air assault operations. Proponents of this concept believe that the MVM concept will provide the joint force with a unique operational maneuver capability to rapidly collapse the cohesion of a regional power under attack by US forces. Since its inception in 1999, the FCS program has undergone some important changes. Of greatest significance is the fact that the weight of the proposed family of vehicles has risen to approximately 30 tons to accommodate new requirements for survivability in the face of a wide range of direct, indirect and mine (IED) threats. Proponents of MVM contend that the heavier combat vehicles should be lifted by a new generation heavy lift VTOL aircraft, the Joint Heavy Lift (JHL) program.[78]

Independent of whether the MVM concept emerges with a fleet of JHL, the FCS armored vehicle program should result in a new generation of "medium-weight" armored vehicles with improved fuel economy. There is a hope that diesel/electric hybrid vehicles may produce a 25% improvement in cross-country fuel economy compared to the current generation of fighting

---

[77] The syllabus at both the National Training Center (NTC) and the Light Infantry Training Center at Ft. Polk has shifted decisively toward the exercise of military skills associated with ongoing counterinsurgency operations in Iraq and Afghanistan.

[78] See BG Robin P. Swan, US Army and LTC Scott R. McMichael, US Army (ret.), "Mounted Vertical Maneuver A Giant Leap Forward in Maneuver and Sustainment," *Military Review*, January-February 2007, for an argument in favor of a national investment in a Joint Heavy Lift (JHL) vertical take-off and landing (VTOL) aircraft to conduct brigade-sized deep aerial assault operations. For a critique of the Mounted Vertical Maneuver (MVM) concept see David E. Johnson, John Gordon IV and Peter A. Wilson, "Air-Mechanization: An Expensive and Fragile Concept," *Military Review*, January-February 2007. The Defense Science Board Task Force on *Future Need for VTOL/STOL Aircraft* explores the daunting cost and technology challenges of successfully developing and procuring a heavy lift VTOL aircraft with the 30-ton payload required to carry a single FCS armored vehicle.

vehicles such as the Bradley family.[79] This new capability is directly relevant to the concepts of operation that will be described in Option C.

Improved Seabasing—The Navy and Marine Corps are making an innovative investment in new seabasing technology. Most important is the development of the Mobile Logistic Platform (MLP). This ship will act as a mother ship for Landing Craft Air Cushions (LCACs) to act as an at sea transshipping node for brigade sized medium and heavy forces. This capability used in conjunction with traditional amphibious assault capability, provides the Navy and Army with a capability to operate several brigades at some depth. Such a capability, however, will rely upon a major build-up of onshore logistics.[80]

Improved Air Sustainment—The Air Force continues to modernize its widebody airlift fleet with the purchase of additional C-17s and the modernization of the older C-5 fleet. A new technology has emerged that will allow the direct delivery of supplies to maneuvering ground forces from these airlifters without requiring the use of a forward airfield. This is the development of the Joint Precision Airdrop System (JPADS). Similar in concept to the JDAM ordnance, the JPADS allows the precision delivery of cargo payloads up to 5 tons from medium altitude (approximately 25,000 feet). Widebody cargo aircraft can provide direct aerial delivery to maneuver forces without landing on austere forward operating bases that may have a very limited capacity or be vulnerable to man portable air defense systems (MANPADS) or long-range rocket and missile fires.[81]

AN OVERVIEW

The irregular warfare challenge posed by the large scale and protracted nature of OEF and OIF has profoundly changed the current national security strategy. Defense planning now focuses on funding and deploying weapons systems and assets designed to support these missions. For example, the costs of these operations have risen from $30 billion in FY02 to $140 billion in FY 07.[82]

The DoD has squeezed budgets of both the Air Force and Navy in order to pay for combat operations by the US Army and Marine Corps. A poster child of these changed priorities is the

---

[79] For a discussion of hybrid powered combat vehicle development programs, see R.M. Ogorkiewicz, "Electric Drives for Combat Vehicles Gain Ground," *Jane's International Defense Review*, May 2004.

[80] Robert W. Button et al, *Warfighting Logistic Support of Joint Forces from the Joint Sea Base* (Santa Monica, CA: RAND Corporation, NDRI MG-649-NAVY, 2007). Recent public reports indicate that the Navy will stretch out the procurement of some of the ships associated with the MPF(F) program to the end of the FYDP due to overall budget pressures and a decision to increase the production rate of the *Virginia*-class SSN. See "FY-09 Budget Delays Prepositioning Vessels, Littoral Warship," *Inside the Navy*, January 14, 2008.

[81] Peter A. Wilson et al, *Sustaining Distributed Deep Operations* (Santa Monica, CA: RAND Arroyo Center, DRR-3875-Army, Forthcoming). For a summary of the JPADS operational experience, see *JPADS Joint Precision Airdrop System, Advanced Concept Technology Demonstration, Joint Military Utility Assessment, Final Report* (Suffolk, VA: US Joint Forces Command, June 2007).

[82] David M. Walker, *DOD Transformation Challenges and Opportunities* (Washington, DC: GAO, GAO-08-323CG, November 29, 2007).

$15 billion-plus crash program to produce and deploy thousand of Mine Resistant Ambush Protected (MRAP) combat vehicles.[83] Even if OEF and OIF wind down expeditiously, both the Army and Marine Corps face massive capital reset costs to repair and replace very large inventories of combat and logistics vehicles.[84] Thus, the Air Force and Navy will face difficulty in sustaining their current programs, which overwhelmingly focus on preparing for traditional major combat operations. As noted below, a return to the pre-Iraq War status quo may not be feasible or desirable, not because of the continued need to meet the challenge of irregular war, but because of the challenge emerging nuclear-armed states.

---

[83] Nathan Hodge, "MRAP Orders Climb to 8,800," *Jane's Defence Weekly*, October 31, 2007. Some estimates suggest that the total buy of the MRAP family of armored vehicles may reach over 20,000 at a cost of some $15 billion. See John M. Donnelly, "Small Wars, Big Changes," *CQ Weekly*, January 28, 2008, for a discussion of the major shift in emphasis by the US Army and Marine Corps away from preparations for high-intensity conventional operations and toward sustained preparations for protracted irregular conflicts.

[84] For a critique of the MRAP crash program, see Andrew F. Krepinevich and Dakota L. Wood, *Of IEDs and MRAPs: Force Protection in Complex Irregular Operations* (Washington, DC: Center for Strategic and Budgetary Assessments, 2007).

VI. "DEFEATING" A NUCLEAR-ARMED REGIONAL POWER: A MODERATE ADAPTATION STRATEGY, OPTION B

Requirements for Option B

The central requirement for Option B is that the United States develops both an advanced counterforce capacity and active defenses that are markedly more effective than either current or programmed capabilities. There are major technological, operational, and organizational challenges to affect such a result.

The Dynamic Regional Nuclear Threat

With only moderate progress made by the United States since the 1991 "SCUD hunt," several key new counterforce capabilities will have to be developed if Option B is to be credibly more effective than our current posture. This challenge is technically and operationally very daunting. Current and future nuclear-armed regional powers will have a wide range of countermeasures to degrade the effectiveness of any counterforce campaign. Some of the more significant are:

- *Upgrading the Missile Threat by Shifting from Liquid to Solid Propellant Booster Rockets* – Following the Cold War technological progression of the United States and Soviet Union, regional opponents have invested in second generation long-range ballistic missiles that rely on solid propellant propulsion.[85] Solid propellant boosters offer the benefits of:

    - Radically reduced launch cycle for a mobile long-range ballistic missile;

    - Radically reduced signature of the deployed missile;

    - Much shorter fly-out time, making boost-phase interception less feasible; and

    - Source of mobile boosters for direct ascent ASAT interceptors.

- *Investing in Substantial Cruise Missile Capability* – Many regional powers will be able to take advantage of the proliferation of cruise missile technology. This will include the

---

[85] Recent public reports suggest than Iran is on the verge of testing a multi-stage, solid propellant MRBM, the *Ashura*, with performance features similar to the Pakistani *Shaheen* II MRBM. The latter has been successfully tested several times and has a range of 2,000-2,500 kilometers. Both MRBMs will operate from a mobile Transporter Erector Launcher (TEL). The former may be compatible with the TELs designed for the liquid fueled Iranian MRBM, *Shahab* 3. See Alon Ben-David, "Iran Adds Ashura to Missile Line-Up," *Jane's Defence Weekly*, December 5, 2007; and Robin Hughes, "Pakistan Test-Fires Hatf 6 Ballistic Missile," *Jane's Defence Weekly*, May 10, 2006, for description of the 2,000 km range, solid propellent MRBM, the *Shaheen* II.

deployment of both long-range subsonic cruise missiles with LO features and shorter range supersonic anti-ship and land attack weapons.[86]

- *Mobility and Concealment* – Foreign observers duly noted the failure of the United States to destroy the Iraqi SRBMs during the first Persian Gulf War. The batting average against them was zero.[87] All Tier One and Tier Two nuclear-armed states have invested heavily in ground mobile ballistic missiles. More recently, several are deploying ground mobile land attack cruise missiles (GLCMs) as well. All have become adept in camouflage, concealment, and deception (CC&D) measures.

- *Hardening and Mobility* – Mobile ballistic and cruise missile systems are being based in hardened facilities. Hardened shelters and tunnels for the missiles and their transporter-erector-launchers (TELs) can be part of a shell game where the attacker is compelled to strike all suspected sites that may have an operational weapon.[88]

- *Super Hardening* – There has been a revolution in tunneling technology. Over the last three decades, a major global industry has emerged in the development and deployment of tunneling systems. There has been no global restraint in the sale of this technology to any willing buyer. Unlike the previous challenges, super hardening places high technological demands on the development of deep to very deep earth penetrating warheads. Further, tunneling technology allows for the construction of deep and dispersed underground facilities that will require precise target identification even if an effective EPW is available.[89]

---

[86] China has made major advances in long-range cruise missile technology that could be exported to a variety of regional powers. See Robert Hewson, "Chinese Air-Launched Cruise Missile Emerges From Shadows," *Jane's Defence Weekly*, January 31, 2007; and Timothy Hu, "CHINA – Marching Forward," *Jane's Defence Weekly*, April 25, 2007. With likely Chinese assistance, both Iran and Pakistan are developing land attack cruise missiles (LACMs). See Farhan Bokhari, "Pakistan Test Fires its Raad Missile." *Jane's Defence Weekly*, September 12, 2007. Not to be outdone, India has a very ambitious joint cruise missile program with the Russian Federation and has developed and deployed a new supersonic cruise missile (SCM), the BrahMos. See Rahul Bedi, "Indian Army Inducts Surface-Launched BrahMos," *Jane's Defence Weekly*, July 4, 2007. The mass deployment of LACMs on wide range of ground, air, and sea launch platforms presents a defender with a new spectrum of threats above and beyond the mere proliferation of ballistic missiles. Also noteworthy is the Chinese development of a modernized version of its H-6 subsonic bomber (a derivative of the 50-year old TU-16 Badger) as a cruise missile carrier. Several nuclear-armed regional powers may acquire a similar capability.

[87] Schneider, "Seeking a Port in the WMD Storm."

[88] With Chinese technical assistance, Iran is developing and deploying a ground launched cruise missile with a land attack capability – a turbojet powered version of the Styx-class anti-ship missile.

[89] For a discussion of the strategic and operational implications of deeply buried facilities see Eric M. Seep, LtCol, USAF, *Deeply Buried Facilities Implications for Military Operations*, Ocassional Paper No. 14 (Maxwell Air Force Base, AL: Center for Strategy and Technology, Air University, May 2000). See also Barbara Demick, "N. Korea's Ace in the Hole," *Los Angeles Times*, November 3, 2003. There are public reports that the North Koreans have provided the Iranian military and security services with technical assistance in the construction of large hardened tunnel facilities. There is a large global market in new and used tunnel boring machines. See the recent edition of *TBM Exchange International*, January 3, 2008 as example of this global industrial activity.

- *Counter-Reconnaissance* – Aside from the extensive employment of CC&D measures, Tier Two regional nuclear powers will attempt to deploy robust air defenses if only to deny the US capability to conduct persistent reconnaissance strike operations over their territory. As noted above, several may develop and/or acquire and deploy a mobile ASAT capability to destroy the US space reconnaissance assets operating in LEO.[90]

In order to make Option B workable, therefore, the United States will need to make radical leaps forward in its ability to locate, track, fix, target, and assess the enemy. Capabilities required to accomplish this very demanding mission include:

Counter-Nuclear Campaign Requirements

- *Persistent and Survivable Reconnaissance* – Today the United States has developed a wide range of persistent reconnaissance means that are not survivable in the face of advanced air defenses. Further, the entire inventory of LEO satellite reconnaissance systems is vulnerable to conventional and nuclear anti-satellite (ASAT) threats. What is needed is a variety of survivable reconnaissance means.

- *Persistent and/or Responsive Strike* – With the small inventory of low-observable B-2 bombers, the United States does not have the capacity to conduct a large scale counter-force campaign that can react to near real-time intelligence of fleeting targets.

- *Effective Munitions* – Although there has been major progress in developing a new generation of munitions, especially those designed to defeat hard targets, the challenge of defeating superhardened and dispersed underground targets remains.

- *Comprehensive SEAD* – To enable persistent and responsive reconnaissance-strike operations, the United States will have to maintain a powerful capacity to conduct SEAD operations against a wide range of aerospace defenses.

- *Survivable Basing* – The reconnaissance-strike and SEAD units must be able to operate from either defended ships at sea or defended bastions located some distance from the likely areas of attack.

- *Operate Over Intra-Theater Distances* – With the exception of North Korea, the large size and geography of current and future regional nuclear powers dictates that all systems designed for the counter-nuclear mission must have substantial range and/or endurance.

- *Trained Joint Forces* – To radically degrade, much less neutralize, a long-range missile force armed with nuclear weapons, the United States military will have to consider the

---

[90] China's January 2007 ASAT test may be a model for other regional powers. The Chinese used a variant of the DF-21 solid propellant IRBM booster to conduct a direct ascent attack on a retired weather satellite with a hit to kill vehicle. See B. Raman, "Anti-Satellite Capability – A Chinese Eye View," *China Monitor*, No. 11, South Asian Analysis Group, January 23, 2007, available at http://www.saag.org/papers22/paper2107.html. Iran and other emerging Tier One and Tier Two powers may acquire advanced air defense systems. See Mark Harrington, "Iran Set to Acquire S-300PTs from Belarus," *International Defense Review*, February 1, 2008.

challenge on the same scale as, for instance, the ongoing effort to defeat the threat of improvised explosive devices (IEDs).

Persistent Reconnaissance-Strike

Naval Unmanned Combat Air System (N-UCAS) – Currently, the Navy has initiated a testing effort to determine whether a Unmanned Combat Air System (N-UCAS) system can operate off of a big deck nuclear powered carrier (CVN) under all-weather conditions. The Northrop Grumman X-47 program is scheduled for such testing by the beginning of the next decade. If the tests are successful then the Navy has the option of making a major investment in the N-UCAS as a long-range reconnaissance-strike platform. Given its expense, it's likely that a major buy will require that the Navy buy fewer F-35C Lightning II's than planned. With a combat radius of 1500 nm, the N-UCAS provides a Carrier Strike Group (CSG) with the capability to conduct deep reconnaissance strike operations. The vehicle's LO features will allow it to conduct sustained operations in the face of sophisticated air defenses.[91]

Pros

- A long range UCAS system based on an aircraft carrier will provide a more survivable persistent reconnaissance and strike capability for the Navy.

- It will partially satisfy the counter-nuclear campaign requirement, especially after a robust SEAD campaign.

- The N-UCAS could be built with robust nuclear hardening requirements.

- Since it operates from a mobile and defended seabase, its base is more survivable than a defended land bastion.

Cons

- Even with LO features the N-UCAS will require an effective SEAD campaign to survive during protracted reconnaissance strike operations.

- The deep reconnaissance-strike mission can be satisfied by further upgrades and a larger purchase of the Tomahawk IV.

- Total operational cost including attrition may be greater than an equivalent F-35C fleet.

- Many of these missions could be satisfied by the F-35C using an F/A-18 buddy tanker.

- The F-35C provides the CSG with much greater mission flexibility, especially to provide close air support and air supremacy.

---

[91] For a more complete discussion of the operational possibilities of the N-UCAS operating from a carrier see Robert O. Work and Thomas P. Ehrhard, *Range, Persistence, Stealth, and Networking: The Case for a Carrier-Based Unmanned Combat Air System* (Washington, DC: Center for Strategic and Budgetary Assessments, 2008).

- Very deep strike missions can be carried out by other means including hypersonic cruise missiles (see below).

A new regional bomber – The DoD has directed the Air Force to develop a new bomber able to operate out of regional bastion by 2018. Current designs call for a subsonic LO vehicle with a payload of approximately 15 tons. Equipped with a next generation AESA radar, this bomber will have the reconnaissance-strike features of a B-2, which is itself being upgraded with an AESA radar.[92]

Pros

- A regional bomber could satisfy the counter-nuclear campaign requirement, especially after a robust SEAD campaign.

- It will be able operate out of a defended bastion.

- With refueling the bomber can act as a persistent reconnaissance-strike platform.

- A new bomber will be able to take advantage of technological advances in LO and improved sensors.

- It can be designed with robust nuclear hardening requirements, for a modest EMP tax, for use against Tier One and Tier Two nuclear threats.

- A manned bomber can carry nuclear weapons under positive control.

- A manned bomber would have the flexibility to conduct a wide range of missions including close air support.

Cons

- A manned regional bomber may not be as survivable as an N-UCAS on a defended seabase.

- A subsonic bomber will not able to respond to fleeting targets from intra-theater distances.

- Very deep strike missions would have to be carried out by other means including hypersonic cruise missiles (see below).

---

[92] See Amy Butler, "Going Nuke, Air Force Secretary Sheds More Definition on Mission of New Bomber," *Aviation Week & Space Technology*, December 3, 2007, for a description of the proposed LO subsonic regional bomber that the Air Force hopes to field by the end of the next decade. The requirement to carry nuclear weapons will likely lead to a manned. The new generation AESA radar systems are now being deployed on a variety of current generation fighter and bomber aircraft. See Martin Steetly, "Raytheon set to Supply AESA Radar for Air Force F-15E," *International Defence Review*, December 1, 2007.

- A fleet of regional bombers may cost more than a fleet of N-UCASs.

Hypersonic Cruise Missiles (HCM) – The Air Force has conducted a series of tests of an air breathing hypersonic vehicle, the HYPERX program. Both the Air Force and Navy have expressed interest in the development of an intra-theater range hypersonic cruise missile. With a speed of greater than Mach 5, these cruise missiles could be launched from a B-52 or warship (either surface or submarine) within the short targeting window that may obtain against mobile long-range missile systems.[93]

Pros

- A Hypersonic Cruise Missile is the logical follow-on to the subsonic Tomahawk.
- An HCM provides survivable strike options from sea-based platforms.
- It would allow the Air Force to sustain the B-52 and/or buy a next generation cruise missile carrier vice the regional bomber.
- It provides a deep strike option for a SEAD campaign.
- An HCM may provide significant EPW capability against hard and deeply buried targets.
- It may also be useful against a wide range of targets including surface ships.

Cons

- An HCM will require a substantial R&D investment to acquire an operational capability.
- Hypersonic flight may still not be fast enough to attack fleeting targets.
- The vehicle may have only a modest payload, which would limit EPW options.
- An HCM will be much more costly than Tomahawk.
- A major HCM purchase may be equivalent to the cost of the N-UCAS or regional bomber fleet.

Boost-Glide Vehicles – DARPA and the Air Force are making a modest effort to develop a boost-glide vehicle, the Common Aerospace Vehicle (CAV). The current concept is to launch the CAV on a small booster, either a converted Intercontinental Ballistic Missile (ICBM) or a

---

[93] Currently, the Hypersonics Flight Demonstration program (HyFly) National Aerospace Initiative is underway to demonstrate a Mach 6+ HCM with a range of 600 nautical miles. See "Hypersonics Flight Demonstration program (HyFly)," at http://www.globalsecurity.org/military/systems/munitions/hyfly.htm.

purpose-built rapid response booster.[94] The Navy has its variant, a boost-glide vehicle launched from a Trident SLBM. The attack vehicle would fly at speeds closer to Mach 20, suborbital velocity, and provide a very responsive strike option.[95]

Pros

- A boost-glide vehicle would provide a more responsive strike option than the HCM.
- It provides for potentially powerful EPW options.
- A CAV would provide SEAD campaigns with a deep strike option.
- It will enable survivable strike from either sea or continental United States (CONUS) bases.

Cons

- A CAV will require a substantial R&D effort to generate an operational capability.
- Each attack vehicle and booster will cost several times that of a HCM.
- If bought in small numbers, the CAV can only be used for uniquely high value targets.
- If bought in large numbers, the total program cost will likely to be similar to either the N-UCAS or regional bomber programs.
- An ICBM-mounted CAV may have arms control treaty issues.

A New Generation of Earth Penetration Warheads

Currently, a 30,000 lb. EPW is under development for delivery by the current generation of penetrating long-range heavy bombers such as the B-2A and B-1B. Smaller payload systems will

---

[94] For an analysis that compares a fleet of new manned long-range bombers with a fleet of boost-glide vehicles, see Russell Shaver, Michael Kennedy and Ted Harshberger, *A Comparison of Manned Bombers vs. Combat Air Vehicles (CAV) in the Execution of Time-Critical Attacks Against Newly Emerging Targets* (Santa Monica, CA: RAND Corporation, DRR-3407-PAF, August 24, 2004). The study acknowledges the critical role of persistent and survivable C4ISR to provide either means of attack with timely targeting information. Overall the study concluded that the CAV or boost-glide vehicle was competitive, if not superior, to a manned bomber acting as a hunter-killer when the number of targets were small and the timeline of holding them at risk was short. Bombers were superior if the number of emerging targets were (relatively) high, their exposure time is short (i.e., less than an hour), and the importance of holding the targets at risk was sufficient to accept the cost of putting the bombers near the target area. Ibid., p.101.

[95] See "X-37 MSP Integrated Tech Testbed," at http://www.globalsecurity.org/space/systems/x-37.htm, for a description of the Pathfinder X-37 Program. This maneuvering re-entry vehicle will be flown on the top of an Atlas V EELV during 2008 and will provide valuable data for the Common Aerospace Vehicle program. See Robie Samanta Roy and David Arthur, "Alternatives for Long-Range Ground Attack" (Washington, DC: CBO, March 2006), and Kosiak, *Arming the Heavens*, for discussions about the costs and benefits of building a large fleet of CAV-type strike vehicles.

have to be developed to provide a similar effect for use on next generation attack systems. A major R&D effort could be made to develop smaller EPW that rely on the kinetic energy provided by either the HCM or CAV-type systems. Also under consideration is the development of a low-yield and very robust nuclear EPW,[96] the Robust Nuclear Earth Penetrator (RNEP). A major challenge in the design of any nuclear EPW is trade-off between depth of penetration, yield, and operational effectiveness. Unlike conventional EPWs, a major concern about the use of a nuclear EPW is whether the combination of depth and yield produces sufficient local and regional fallout to gain an assured target kill. Given that a deeply buried target may be an extensive tunnel complex, multiple nuclear strikes of considerable yield may be required to cripple or destroy the underground complex in question. Obviously, a new generation nuclear EPW beyond the current B-61-11 aerial bomb would provide a wider range of options, but the use of nuclear explosives may prove to be an unworkable solution to the limits of large, non-nuclear EPW such as the soon to be operational air delivered Massive Ordnance Penetrator (MOP).

Resurrecting Joint Counter-Nuclear Campaign Training

The joint force with the Air Force in the lead will have to resurrect the "Rolling Sands" counterforce and active defense exercises of the 1990s. The demands of the current operations in Iraq and Afghanistan have forced cutbacks in training not directly related to those ongoing major conflicts. Looking beyond the major US military presence in Iraq, however, a new generation of counter-nuclear exercises should be initiated.

Active Defenses

The current inadequacies of US and allied missile defenses are well known. These substantial gaps will need to be closed if Option B and especially Option C are to be plausible.

There are steps that can be taken to further improve US and its allies' air and missile defense capability. They include the following possible initiatives:

Major Procurement of Aerospace Defenses – Current generation missile and air defense systems should be deployed in very large numbers. Furthermore, those systems should be modified as necessary to ensure that they can operate in an electronic environment disrupted by EMP (see below).[97]

Pros

---

[96] At this time, Congress has zeroed out the funding to develop the Robust Nuclear Earth Penetrator (RNEP) as follow-on the current nuclear EPW, the B61-11. Currently, a major effort is underway to develop and deploy a 15-ton conventional EPW that is compatible with the B-2A. See Sweetman, "A MOP For Those Tough Clean-Up Jobs."

[97] For example, the joint US and Japanese next generation SM-3 interceptor could have more robust EMP requirements levied on the design. Currently, the plan is to deploy the new SM-3 interceptor with a larger diameter flight motor that will allow the deployment of a larger and more capable hit-to-kill vehicle. This next generation SM-3 is a candidate for major procurement during the middle part of the next decade.

- Major procurement of current interceptors such as PAC-3 and THAAD addresses the problem of saturation attacks by regional opponents armed with large numbers of SR/MRBMs.

- Air defenses will have to be much more robust to deal with mass cruise missile attacks.

Cons

- Major procurement of contemporary aerospace defense systems will lock the United States into systems that are vulnerable to next generation countermeasures and the use of nuclear weapons to generate EMP.

- It will also limit the R&D and procurement funds available for the next generation of active defenses.

Major Procurement of Naval Aerospace Defenses - This is the rationale for the SM-3 interceptors on board Aegis-equipped cruisers and destroyers. A strong case can be made that the next generation cruiser should be optimized for the ballistic missile defense mission.[98]

Pros

- The Navy will have to step up to the joint force requirement to provide robust and survivable aerospace defenses – the Navy budget for SM-3 procurement is far too low. The jointly developed and more capable US-Japanese Block II variant of the SM-3 would be the appropriate version for mass production.

- The next generation guided missile cruiser should be optimized as a mobile BMD system. This might be a hull derived from an amphibious ship to provide the volume and weight needed for a very large number of interceptors and power for missile defense radars.

Cons

- The next generation guided missile cruiser optimized for the BMD mission may not be as flexible as alternative thin-hulled and higher speed designs.

- The Navy runs the risk of developing an unaffordable next generation cruiser.

---

[98] There are proposals within the DoN to develop a next generation cruiser using the amphibious *San Antonio*-class (LPD-17) hull. The argument is that a battle cruiser optimized for the missile defense mission would not need the high sprint speed of a more traditional slender hull design. A ship of this size, some 30,000 tons, could carry a very large inventory of stand-off weapons including the next generation SM-3. Further, the ship would have the power, volume, and displacement to deploy very powerful X-band radar and a number of shorter range directed-energy weapons. Powerful members of the Congress are now advocating that this next generation crusier be nuclear powered. A warship with these features would cost in the neighborhood of $6 billion vice the proposed cost range of a smaller, conventionally powered design of some $3-4 billion.

Major Investment in Boost-Phase Intercept – If successfully developed via the NCADE program, there is the opportunity of converting both the F-22 and F-35 to boost-phase interceptor platforms. This is an opportunity to take full advantage of the massive investment in two LO fighter aircraft fleets, both of which are capable of fighting for air supremacy deep inside hostile airspace.[99] In essence, LO tactical aircraft would conduct both counterforce and boost-phase intercept operations over suspected mobile missile launch areas.

Pros

- The LO attributes of the F-22 and F-35 can be fully exploited to maintain defensive combat air patrols (CAP) over suspected mobile ballistic and cruise missile launch areas that are protected by robust air defenses.

- This CAP could be sustained by the use of aerial tankers outside of the hostile airspace.

- This would provide a strategically and operationally agile means of reinforcing local ground bases and Navy off-shore BMD capability.

Cons

- This concept of an anti-missile CAP may be feasible only for smaller countries such as the Democratic People's Republic of Korea (DPRK). Larger countries such as Iran will require a very large and possibly unsustainable number of aircraft over a wide range of suspected launch sites located deep (more than 700 miles) in their territory.

- Even with LO features, conducting survivable, protracted reconnaissance strike operations will require a successful SEAD campaign – a challenge similar to deep N-UCAS operations.

- Aircraft conducting deep, boost-phase anti-missile operations will not be available to provide either a boost-phase or mid-course defense against cruise missile attacks.

More Robust C4ISR

Defend the NSS Infrastructure at MEO and GEO – The critical role of GPS providing PNT at MEO and the Extremely High Frequency (EHF) communication constellation at GEO suggests a possible response to a robust non-nuclear and nuclear ASAT threat. Satellites operating a LEO may be too vulnerable to mobile direct ascent ASATs and directed energy weapons to be defended in a cost effective manner. High altitude satellites can only be directly attacked with non-nuclear or nuclear ASATs launched by larger ICBM-class boosters. Tier One and Tier Two opponents are likely to have a limited high altitude space launch capability. Any high powered directed energy weapon facility is likely to be detectable and subject to precision air or missile attack. This suggests that a combination of counterforce attacks on ASAT launchers and/or the

---

[99] See Robin Hughes, "Paris Air Show: Raytheon Prepares to Flight Test NCADE Interceptor," *Jane's Defence Weekly*, June 27, 2007.

use of airborne boost-phase interceptors would provide a robust defense of these important NSS assets. Further, the next generation PNT and communication satellites could be provided with additional propulsion and nuclear hardening to limit the effectiveness of any direct ascent nuclear-armed ASAT that survives and aerial interdiction campaign.

High Altitude Air Ships – Wide aperture Ultra High Frequency (UHF) radars could be flown on unmanned airships that operate at 60,000 feet and provide wide area surveillance coverage. With a solar power and fuel cell combination, these airships may have an endurance measured in weeks.[100]

Pros

- A wide-aperture low frequency radar is optimized to detect LO aerial platforms.

- A high altitude airship can act as a pseudo-satellite to provide wide area surveillance and/or communication relays.

- It can provide a back-up for the loss of NSS services.

Cons

- The technology requires a major R&D effort to create an operational capability.

- At very high altitudes (above 70,000 feet), the airship will have a very small useable payload.

- At lower altitudes (around 60,000 feet), the airship will have to deal with high altitude winds.

- The enhanced C4ISR mission can be satisfied by long-endurance, high-aspect-ratio winged UASs such as Global Hawk.

Space Based Interceptors (SBIs) – A constellation of LEO satellites could be deployed as part of a BMD system. These satellites could be equipped with either hit-to-kill (HTK) interceptors or high energy laser (HEL) systems.[101]

---

[100] A RAND Arroyo Center study notes that a high altitude airship will have a very modest payload when operating above 60,000 feet. To carry a payload of 500 pounds to an altitude of 70,000 feet will require an airship over 500 feet long and capable of holding 5 milion cubic feet of helium.

[101] See Vickers and Martinage, *Revolution in War*; and Bob Preston et al, *Space Weapons, Earth Wars* (Santa Monica, CA: RAND Corporation, 2002), for a cost-benefit analysis of a constellation of space based weapons using either hit-to-kill (HTK) vehicles or High Energy Lasers (HEL). These studies conclude that a space-based defense constellation is very inefficient since only a small number of space defense vehicles are in view of a potential ballistic missile launch area at any given time. Those few space defense vehicles in view can then be overwhelmed by salvo launches. Further, they note that the defense vehicles themselves can be attacked by direct ascent ASATs.

Pros

- SBIs provide boost-phase interception of the full range of ballistic missiles.
- They provide a means to defeat exo-atmospheric decoys.
- SBIs have an inherent ASAT capability.

Cons

- It will require a very large R&D effort to develop either a HTK or High Energy Laser (HEL) satellite.
- The difficulties in the ABL program prove the daunting R&D challenge of developing a HEL suitable for space launch and provisioning for multiple shots.
- BMD satellite constellations in LEO are very inefficient. To gain continuous coverage against a suspected launch site, the constellation must maintain a very large number of satellites on orbit but out of view.
- A BMD constellation in LEO is vulnerable to non-nuclear and nuclear ASATs.
- A LEO hit-to-kill interceptor requires the development of a very fast and light weight kinetic energy kill vehicle (KKV). The deployment of solid propellant boosters doubles the velocity requirement of the KKV.
- A large LEO BMD constellation will cost as much as a significant regional bomber program, approximately $50-100 billion.

Preparing for EMP and HAND Attacks

These new long-range strike and active defense capabilities will have to operate in an environment where one or more nuclear weapons may have been detonated to generate wide area electromagnetic effects or detonated at a high altitude to "pump" the van Allen belt to degrade all low earth orbit satellites. As for new counterforce means, all next generation active defenses will have to be resilient in the face of a limited nuclear weapons use as part of an opponent's counter reconnaissance campaign. First generation fission weapons will generate modest EMP effects. On the other hand, when detonated at the appropriate altitude, even a first generation nuclear weapon of 20 kilotons will energize the van Allen belt causing a radical reduction in the operational lifetime of any unhardened satellites operating in LEO.

As noted above, high priority PNT, surveillance, and communication satellites operating at MEO and GEO can be built with enhanced nuclear hardening specifications. This would reduce the chance that a single nuclear-armed ASAT could achieve multiple kills of these higher altitude NSS assets.

This vulnerability will require considerable reflection upon the best tradeoff between attempting to defend a vulnerable satellite constellation versus more investment-intensive, resilient "air-breather" reconnaissance capability. In any case, the United States will need to have redundancy in its reconnaissance capability in order to provide C4ISR even in the face of either expected or unexpected counter-reconnaissance attacks.

On Ground Force Operations

Given the realities of a nuclear environment and the vulnerability of land forces to nuclear strikes, their role will be highly circumscribed. As discussed in Chapter 4, the history of ground forces in a nuclear environment suggests that their effectiveness is starkly limited. The US will not be able to mass large ground forces and their associated logistics train within easy range of an adversary's regional long-range strike forces. This contradicts the combat logistics requirements of high intensity conventional ground force operations. Just supporting a single heavy division equipped with four BCTs, a fires brigade, and an aviation brigade during five days of heavy combat will require approximately 1,500 logistics and escort vehicles operating over a distance of 500 miles from a large theater support base.[102]

Operating from Stand-off Distances

There are two broad choices for operating ground forces in the shadow of direct nuclear attack. The prospect of nuclear strikes demands that ground forces be highly mobile, dispersible, flexible, and resilient (armored). They must be able to quickly mass fires and avoid presenting an attractive target to enemy attacks. Thus US ground forces will have to either operate from a very heavily defended bastion some distance from the tactical theater of operations or they will have to operate from a heavily defended seabase some distance off-shore – or both.

Specialized Small Ground Units – In either case, this requirement suggests that ground forces used in a counter-nuclear campaign will be highly specialized and relatively small in number – no more than few brigades. An especially important mission of these special operations-capable ground units will be to conduct deep raids in support of the counter-nuclear campaign and to provide targeting and reconnaissance capabilities for stand-off platforms.

New Model Airborne Forces – A new design for the Army airborne forces is needed to conduct deep maneuver and/or support special operations. It will have be motorized with light armored vehicles and delivered from medium altitude via Joint Precision Airdrop System (JPADS) technology. *Ranger*-type units can be trained to conduct combat airborne assault from cargo aircraft operating at medium altitude. Using the JPADS technology described above, it is conceivable that combat personnel mounted in small all-terrain and/or light armored vehicles could conduct a form of MVM against selected high value targets.[103]

---

[102] See Wilson, *Sustaining Distributed Deep Operations*.

[103] The Air Force and SOF communities have worked out the training procedures for dropping cargo and personnel from medium altitude (approximately 25,000 feet). As the cargo aircraft decompresses, all personnel will have to use oxygen and wear environmental protection due to the extreme cold. C-17s were used repeatedly to drop humanitarian relief supplies during the opening weeks of Operation Enduring Freedom without any adverse health incidents among the Air Force flight crews.

Fighting and Sustaining from a Bastion – A bastion will likely be one to two thousand miles from the theater in question. Its facilities will have to be heavily defended from a wide range of air and naval threats. (A heavily defended Guam or Diego Garcia are two examples.) Therefore the ground maneuver elements will have to be supported by aviation to provide intra-theater maneuver lift, fire support, and sustainment.

As noted above, any counter-nuclear raiders may be required to seize an airfield for the ingress of further reinforcements and supplies and not only for their aerial egress. At present, only widebody airlifters such as the C-17 and C-5 aircraft can sustain brigade-sized forces over intra-theater distances. This means that an airfield that has been seized by either airborne and/or amphibious assault forces will be very vulnerable to a direct nuclear attack. To avoid the use of a single fixed airfield, some form of aerial support is needed.

In the near-term, it will be possible to use the JPADS to provide support for maneuvering units without relying on a large airfield. Supplies can be directly delivered to the maneuver forces using the airlift fleet while it operates at medium altitude.[104]

Operating from the Seabase – The Navy and Marine Corps are making an innovative investment in new seabasing technology as described previously. These sea-based forces could provide support to any airborne operation. As an alternative to seizing an airfield for egress of airborne units, the seabase may be used to facilitate the egress of the entire the intervention force via a line of communication to the shoreline seized by Marine units. The ships of the seabase will have to be protected from a wide range of ballistic/cruise missiles and submarine delivered nuclear threats.

SUMMATION

Option B is based upon the proposition that a new type of "Flexible Response" strategy is sufficient to deal with emerging regional nuclear-armed powers. The strategy posits that the main purpose of a future US military response to any major act of aggression by a nuclear-armed state is to deny success rather than to defeat the regime. On the other hand, the high performance counter-nuclear campaign capability does not preclude a more traditional regime change operation that relies on large scale offensive ground combat operations by the United States and willing regional and international allies. The key to any regime change operation will be the success or failure of the counter-nuclear campaign in disarming the nuclear threat. Without its forceful disarmament, any US administration is likely to accept a regional war outcome of limited strategic aims vice total defeat of the nuclear-armed adversary.

After an opponent has attempted to use its nuclear forces to facilitate a major act or acts of aggression, a stalemate may be unacceptable to the United States. This strategic deficiency suggests that a more robust capability than that described in Option B will be needed to conduct

---

[104] JPADS technology provides the Army with the opportunity to consider the option of conducting air assaults from medium altitudes. With proper training and equipment, combat units could conduct air assaults using precision airdrop technology.

military operations against a regional nuclear power in order to achieve victory rather than a stalemate.

VII. AN AGGRESSIVE ADAPTATION STRATEGY OF NEUTRALIZATION AND DEFEAT: OPTION C

On the Conduct of a Counter-Nuclear Campaign

The central premise of this strategy is that the United States can develop a counter-nuclear campaign capability that is an order of magnitude more effective than the current and programmed efforts. There are major technological, operational, and organizational challenges to realizing such a result.

On Counterforce

The combined effect of the broad advances in counterforce as described in Option A allow the US to readily attack and destroy all known targets on the surface of the earth, land or sea. With its air supremacy, the US military can readily defeat conventionally equipped and organized armed forces under all weather conditions. What these major advances in combat capabilities have not provided is military dominance against insurgent forces or nation states armed with long-range ballistic and cruise missiles. To defeat both classes of threats, the United States will have to make a sustained investment in a new generation counterforce capability as described in Option B.

On Active Defenses

The Option C investment strategy is more robust than that of Option B. Beyond the aerospace defense postulated in Option B, there would be the need to develop a boost-phase interceptor platform that could survive in the face of robust air defenses. Further, there is the possibility of arming the next generation AWACS with a lookdown-shootdown capability, optimized to destroy LO cruise missiles and S/MRBMs. This armed air defense vehicle would have to operate from a distant defended bastion.

Building a robust C4ISR infrastructure

As indicated previously, a nuclear-armed regional opponent may conduct a vigorous counter-reconnaissance campaign against the US LEO satellite infrastructure. This could include the use of non-nuclear (hit to kill) or nuclear-armed interceptors. A less technologically developed opponent might choose to use nuclear armed interceptors to guarantee a kill. The use of these nuclear interceptors by accident or design will cause the energizing of the lower van Allen belt and put at risk all satellites operating in LEO.

Attacks against satellites operating at higher altitude, most specifically the Navstar GPS constellation operating at MEO and EHF communication satellites at GEO by nuclear armed interceptors using IRBM/ICBM rocket boosters are plausible. As noted previously, this higher altitude region of the NSS services infrastructure may be defended by a mixture of counterforce, boost-phase intercept and more survivable satellites. Those defenses may still prove inadequate, however.

This prospect highlights the need to develop a very robust C4ISR system of systems that can operate with important elements of the National Security Space (NSS) infrastructure seriously damaged or destroyed. Several alternative strategies should be considered:

Rapid Satellite Reconstitution – A fleet of small reconnaissance, navigation and communication satellites could be built and held as ready-to-launch replacements for the degraded/destroyed LEO "peacetime" infrastructure. These smaller satellites would use a rapid launch capability from either a smaller, next generation launcher or the current fleet of ICBMs and SLBMs. This is the essence of the Operationally Responsive Space (ORS) concept.

Reliance on Non-Satellite Means – The ORS concept may not prove feasible for cost and effectiveness reasons. The clear alternative is to more heavily rely upon air-breathing means to provide surrogate intelligence, navigation, and communication services. This could include a major investment in two types of unmanned air systems. The first is a Global Hawk or HAA-class UAS design to provide theater forces with wide-area communications and to act as a surrogate for the degraded GPS constellation. The second is a UAS designed to conduct deep reconnaissance in hostile airspace. This might be a variant of the proposed N-UCAS that is optimized for reconnaissance. It can provide targeting information to rapid response weapons such as HCM or a boost-glide vehicle launched from either a submarine or CONUS.

On EMP and HAND "Taxes"

The new long-range strike and active defense capabilities will have to operate in an environment where one or more nuclear weapons may have been detonated at a high altitude to "pump" the van Allen belt and degrade satellites in LEO or to generate wide area EMP. Therefore, all next generation active defenses will have to be resilient to the limited use of nuclear weapons as part of an opponent's counter reconnaissance campaign.

Option C requires that all future counterforce, active defense, C4ISR, and maneuver forces be modernized with systems with enhanced resilience to EMP. The EMP "tax" will range one to five percent of the total system cost. How much EMP hardening is enough? The answer will depend upon an assumption of the size and composition of the nuclear challenge faced US forces.

Large Scale Follow-on Ground Operations

A key assumption of this option is that the United States may have to forward deploy large ground forces in the theater of operation before the completion of a "successful" counter-nuclear campaign. Success would be defined as the assured destruction of at least 90% of the opponent's nuclear arsenal. This campaign might take several weeks of combined air and missile and deep special operations missions. Even after the completion of this campaign, a decision to conduct a large scale military operation against the adversary with the objective of overthrowing and replacing the regime would remain risky. The expeditionary force would have to plan on the prospect that at least a handful of nuclear weapons survived and could be used as part of the adversary's counter-invasion plan. This suggests that all of the US ground maneuver forces engaged in this follow-on operation would have to be equipped with heavy and medium-weight armored fighting vehicles. Armor will provide substantial protection from the blast, heat, and radiation effects of first generation fission weapons.[105] Most likely, the residual nuclear arsenal would be used via hidden atomic demolition munitions, suicide vehicles, or short-range rockets from hidden and mobile launchers.

In the event that one or more nuclear weapons are detonated on the ground, then the field forces will have to operate in a battlefield that is contaminated with dangerous, if not lethal, doses of radiation. The maneuver forces will have to be prepared to operate "buttoned up" in their armored vehicles—if only to maneuver out of the zones of highest radioactive contamination.

A NEW MODEL ARMY AND MARINE CORPS

If conducting major theater operations under nuclear attack becomes a central planning requirement, both the Army and Marine Corps will have to consider a major restructuring and a radically revised concept of operation for their ground maneuver forces.

Operating from Stand-off Distances – There are two choices for conducting significant ground operations in the shadow of direct nuclear attack. They will have to either operate from a very heavily defended bastion some distance from the tactical theater of operations, or they will have to operate from a heavily defended seabase some distance off-shore.

Sustaining from a Bastion – At present, only widebody airlifters such as the C-17 and C-5 can sustain brigade-sized forces over intra-theater distances. This requires that either airborne and/or amphibious assault forces seize an airfield. That airfield will then be vulnerable to a direct nuclear attack. To avoid the use of a single vulnerable airfield, some form of aerial support is needed. In the near-term, it may be possible to use the JPADS to provide support for maneuvering units without relying on a large airfield. Supplies can be delivered directly to the maneuver forces using the airlift fleet operating at approximately 25,000 feet, safely above most ground threats. Alternative concepts discussed below include the development of a fleet of Joint Heavy Lift (JHL) vertical take-off and landing (VTOL) cargo assault aircraft or the development of a fleet of semi-buoyant air ships (SBAS).

Operating from the Seabase – As noted, the Navy and Marine Corps are making an innovative investment in new seabasing technology. These sea-based forces will provide logistics support to large scale ground operations. The ships of the seabase will have to be protected from a wide range of ballistic and cruise missiles and submarine delivered nuclear threats.

On Heavy, Medium, and Light Brigade Combat Teams (BCTs)

Currently, the US Army has a mix of heavy (armored/mechanized infantry), medium (Stryker), and light (infantry) maneuver units organized into modular Brigade Combat Teams (BCTs). The BCT has become the basic unit to provide close combat power and will be attached during wartime to a division headquarters. Drawing on its experience in Iraq and to a lesser degree in Afghanistan, the Army will attach a specific mix of BCTs to a division headquarters based on local operational circumstances.

---

[105] The lethal bast, prompt radiation, and thermal effects radius of a low yield nuclear weapon will be reduced by a factor of approximately five for troops operating under armor protection vice being in "soft" vehicles.

43

A Bias in Favor of Armored Mobility - While conducting military operations under the threat or actual use of nuclear weapons, the most plausible force mix will be comprised of heavy and medium BCTs. Both will have the inherent armor protection and mobility to more effectively survive and function after a first generation nuclear weapon detonation. Infantry BCTs, if fully equipped with MRAPs, will be as survivable as a Stryker brigade. All BCTs will have organic logistics and engineering elements equipped with armored protection. The current crash program to produce thousands of MRAP vehicles in response to the Iraqi and Afghani IED threat can be leveraged. As the Iraq War winds down, the MRAP inventory could be used to replace the large inventory of soft vehicles currently deployed with logistic and engineering units at all levels. There will be no role for "soft vehicles" within the BCTs and division units that are conducting operations under a nuclear threat.

"Going to Ground" – With an organic combat engineering capability, the BCTs, and especially the division and above echelons will have to rapidly build field fortifications to mitigate the effects of nuclear weapons use. As noted, any in-theater defended bastions will have to be configured in a locally dispersed and hardened fashion.[106]

Dispersed Operations - BCTs will have to be prepared to operate tens of miles from their neighboring BCT. The division echelon will basically be a C4ISR node for multi-BCT operations and will have little organic fires and logistics capability other than a combat engineering capability to "go to ground." Both the BCTs and division echelon will rely heavily on non-organic assets to provide ISR, long-range fires, combat aviation, aerospace defenses and logistics support.

Reach Back – The traditional forms of organic combat support and combat service support will have to be radically pared back to reduce the overall vulnerability of the ground maneuver units to nuclear strikes. Most of these assets will have to be delivered by air from heavily protected bastions or seabases. A key requirement will be the need to deliver all consumables, especially water and Petroleum, Oils and Lubricants (POL), by aerial means.

Reliance on Guided Munitions – All indirect fire systems whether organic to the BCT or provided by outside sources will rely heavily on guided munitions. Their use will radically reduce the logistics throughput of these very logistically intensive elements of the ground maneuver units.

This approach is not dissimilar to the Army's Future Combat Systems (FCS) modernization effort, which is aimed at building more combat-capable BCTs that can operate in a dispersed and tactically agile manner. The Marine Corps will have to reconfigure in a somewhat similar fashion to be able to operate in a nuclear shadowed environment. The Marine Expeditionary Brigade will have to be configured so that the combat maneuver element can operate without significant portions of its combat support element, which will have to stay off shore to be protected.

---

[106] Extensive US, Soviet, and Chinese nuclear exercises during the Cold War highlighted the value of field expedient fortifications to reduce the lethal radius of low-yield nuclear weapons.

Refocusing the FCS Program

The central rationale for this Army modernization program is no longer just to be able to more efficiently defeat a conventional opponent. The new rationale for the FCS program should be to modernize the Army so that it can credibly conduct large scale ground operations against a nuclear armed opponent. The 30 ton family of fighting vehicles will be equipped with more fuel efficient hybrid engines. In a nuclear environment, their heavier weight, lower profile and ballistic shape will make them much more survivable than the widely deployed MRAP family of armored vehicles. The latter, with its much high center of gravity, will be more vulnerable to nuclear weapon blast effects. Thus, it may make sense for the FCS to replace portions of the Bradley family of vehicles while using MRAPs to replace soft service and support vehicles.

The key enabler of the FCS program is the high bandwidth communications coupled to battlefield and theater surveillance systems that provide a "first shot, first kill" capability using stand-off guided munitions. If deployed Army-wide, the enhanced C4ISR suite derived from the FCS program will be consistent with the attributes described above to create units more capable of operating in a nuclear shadowed environment. This C4ISR suite will have be hardened to withstand the EMP effects associated with second generation high yield nuclear weapons. This calls for an EMP "tax" higher than Option B since involves the modernization of the entire ground force. Further, Option C assumes that future operational threats may use low or high yield second generation nuclear weapons—if only to be used as a wide area generator of EMP effects.

On Vertical Maneuver Operations

Historically, advocates of vertical maneuver believed that the extensive use of helicopters or more advanced VTOL aircraft could address the nuclear threat problem. Although the Army has a large inventory of helicopters to provide combat support to the ground maneuver elements, these modular aviation brigades are very vulnerable to direct nuclear attack. The aircraft are very short range, require a large POL supply line, and are very vulnerable due to the "softness" of the air and ground support vehicles to nuclear weapons effects.[107]

The Joint Heavy Lift (JHL) Concept

A new generation of very heavy lift VTOL aircraft could be developed after 2020. These VTOL aircraft would have a payload of some 30 tons and an unrefueled combat radius of 500 miles.

Pros

- With their VTOL capability, the assault cargo aircraft can operate from a wide range of dispersed airfields thereby greatly increasing the targeting problem of the regional opponent.

- FCS equipped BCTs can be maneuvered through the air to conduct the equivalent of a 21st air assault.

---

[107] Swan and McMichael, "Mounted Vertical Maneuver A Giant Leap Forward in Maneuver and Sustainment"; and Johnson, Gordon IV and Wilson, "Air-Mechanization: An Expensive and Fragile Concept."

Cons

- Approaching any landing zone, these large aircraft will be severely threatened by local mobile and hidden air defenses.

- Local air defenses must be suppressed by the SEAD campaign associated with the counter-nuclear campaign.

- Each JHL will likely cost more than $200 million and require a decade-long R&D effort totaling some $10-15 billion.

- To carry out a BCT-sized aerial assault will require the procurement of several hundred air vehicles at a total system cost similar to the current FCS program, some $200 billion plus.

- Similar to widebody airlifters such as the C-17 or European A-400M, the aircraft cannot be provided hardened shelters; therefore, they will be able survive only if operating from a heavily defended bastion located within 500 miles of the ground maneuver forces.

- Since fixed-wing aircraft such as the C-17 will be able to operate efficiently from a distance of some 1500 miles from the theater of operation, they will have an inherently more defensible bastion.

This suggests that the Army should not invest in a high performance VTOL aircraft, but rather train and optimize the medium and heavy ground maneuver BCTs to be able to operate at considerable distances, with only Air Force long-range aviation sustainment support. As noted, the Air Force airlifters can provide direct re-supply to maneuvering BCTs and/or an austere division echelon via the use of JPADS.

Precision Air Drop

As discussed above, if the technology of JPADS could be exploited to develop precision air drop for payloads weighing up to 10 to 15 tons, it could lead to a different type of MVM with modernized airborne rather than VTOL air assault forces.[108]

Pros

- It allows the intra-theater air assault of a BCT-sized force to seize a key strategic target.

---

[108] During 2007, the two different JPADS concepts have been used operationally in Afghanistan and Iraq. See Staff Sgt. Carlos Diaz, US Central Command Air Forces, "First JPADS Airdrop over Iraq a Success," http://www.mnf-iraq.com/index.php?option=com_content&task=view&id=10071&Itemid=128, February 20, 2007; and authors' conversation with Rich Benney, Soldier's Systems Development Command, Natick, Massachusetts. Benny indicates that, with funding, a 5-ton JPADS could be operational by early next decade. With this capability, it is possible to consider the precision aerial delivery of light combat vehicles, either armored HMMWVs or their follow-ons, from medium altitude.

- Although smaller than the FCS family of combat vehicles, this new class of armored vehicle will provide the airborne forces with significantly enhanced protected mobility and firepower.

- Regular infantry can be trained to conduct air assault operations from medium altitude. This would allow the Air Force to over-fly local air defenses, bypassing a very difficult threat to destroy through a SEAD campaign.

- JPADS technology is maturing rapidly and has been used operationally in both Iraq and Afghanistan.

The cost of developing and deploying a BCT-sized airborne capability is a fraction of the cost of a BCT-sized JHL-enabled air assault unit.

Cons

- Airborne assault is too risky from medium altitude. Only Ranger-type units can be reliably trained to avoid decompression sickness while conducting high altitude low opening (HALO) operations.

- The airborne community will never accept the concept that airborne infantry ride down in a combat vehicle.[109]

- The smaller combat vehicles associated with the MVM concept are not as survivable as the larger and more capable FCS family of combat vehicles.

The Semi-Buoyant Air Ship

Lockheed Martin has been conducting experimental flights of a small scale, semi-buoyant airship (SBAS). The vehicle relies on a combination of buoyancy and forward flight speed to fly at low-altitudes as a powered airship. This concept could provide ground forces with the ability to deliver by air several hundred tons of supplies and combat vehicles. The unrefueled radius of the SBAS is greater than either the C-17 or C-5 airlifter and with a landing area no larger than a 1,500 foot radius circle; the SBAS could bypass port facilities and provide direct delivery to maneuvering combat units.[110]

Pros

---

[109] It is noteworthy that Soviet airborne forces were equipped with light armored vehicles, the BMD series, and planned airdrops with the vehicles manned with combat personnel.

[110] See Wilson, *Sustaining Distributed Deep Operations*; and Christopher Bolkcom, *Potential Military Use of Airships and Aerostats* (Washington, DC: CRS, RS21886, May 9, 2005). Lockheed Martin flew a small-scale prototype of the hybrid air ship, or SBAS, during the winter of 2006. DARPA had planned to fund a full-scale prototype as part of an ACTD, but that program was cancelled during the summer of 2006. An alternative SBAS concept is the SkyCat 1000. See a description of the "SkyCat 1000," at http://www.globalsecurity.org/military/systems/aircraft/skycat.htm.

- The development costs of a 500-ton payload SBAS are one or two billion dollars.

- A small fleet of 10-20 airships could support multi-division operations from a defended theater bastion.

- The SBAS will be able to over-fly ports and other restricted land lines of communication that may be lucrative nuclear targets.

- The SBAS will be very resilient to small arms ground fire.

Cons

- With a full payload, the maximum cruising altitude of a large SBAS is 12,000 feet. This limits the SBAS to low-lying terrain, which will exclude much of the world.

- When operating with a full payload, the SBAS will be required to take on ballast (water) after off loading its cargo.

- The airship does remain vulnerable to high rates of fire from small and medium caliber anti-aircraft artillery (AAA).

- Ownership, and therefore service and support, of a fleet of SBAS remain uncertain.

Expanding Seabasing

Currently the Navy plans to deploy only one upgraded Maritime Prepositioning Force (Future) (MPF (F)) by the middle of the next decade. A strong case can be made that this investment should be accelerated and expanded to include two additional unit sets with additional capacity to support multi-brigade operations of both the Marine Corps and Army. Alternatively, the MLP program could be expanded independently of the number of MPF (F) squadrons to facilitate a larger joint off-shore operational capability.

The Marine Corps VTOL fleet will be more survivable than forward deployed JHL aircraft since it will be able to operate from a protected seabase. On the other hand, the Marine Corps decision to acquire the CH-53K as its primary VTOL cargo aircraft suggests that a future seabase will have to operate no more than 50 miles off-shore in support of either an amphibious or administrative operation. Although the V-22 tilt-rotor aircraft has a longer combat radius than the CH-53K, it can carry only one third the payload of the larger, but slower, helicopter.

Defending the Bastions and Seabase – The major in-theater bastions and seabases will have to be heavily protected. The carrier battlegroup will provide sufficient air defenses. A robust sea-based ballistic missile defense capability will be needed. This is the rationale for the SM-3 interceptors on board Aegis-equipped cruisers and destroyers. A case can be made that the next generation cruiser should be optimized for the ballistic missile defense mission. Further, the theater bastions and seabases will have to be protected against attacks by submarines armed with either long-range, nuclear-tipped cruise missiles or nuclear-armed torpedoes.

Protecting Regional Allies

Given the limits of even a robust counter-nuclear campaign and the prospect that US and allied military forces may have plausible protection options, a future nuclear-armed opponent may choose to threaten the major cities of US allies within range of its nuclear strike systems. This is an acute dilemma in several possible theaters of operation (e.g., Northeast Asia and the Persian Gulf, where major population centers lay within several tens or at most a few hundred miles of likely nuclear launch sites). Even Option C only assumes that those cities will be protected as much as possible by the counter-nuclear campaign. That may not be good enough for the regional political leadership; therefore, the threat of retaliation will have to come into play.

Not surprisingly, a regime's willingness to consider "city busting" attacks will likely be contingent upon whether the regime believes itself to be in mortal risk. This is the inherent strategic dilemma that US military planners and the senior political command authority will have to consider. Yet, it may be critical, if only as a deterrent posture, that the United States have a militarily credible means of destroying the regime without destroying the nuclear armed state, much less the region of interest. A central rationale for Option C is to provide just such a capability that is beyond the simple threat of retribution.

VIII. COUNTER-NUCLEAR CAMPAIGN OPTIONS: WHAT APPEARS FEASIBLE?

Nascent or Tier One Capabilities: Limited Retaliatory Capability

The emerging Tier One nuclear arsenal will be small in number (less than two dozen) and consist of first generation fission weapons with yields of no more than 20 kilotons. The primary means of delivery will be either jet fighter aircraft or first generation liquid propellant ballistic missiles. North Korean currently possesses, and Iran is trying to acquire this type of capability. The role of this arsenal will primarily be to protect the regime from an outright military overthrow.

Demonstrations – This could range from an underground nuclear weapon test, not unlike the DPRK test in October 2006, to a high altitude detonation designed to demonstrate an understanding of the EMP/HAND phenomenon. Such demonstrations would be designed to deter and/or coerce neighboring powers that are allied with the United States from taking or supporting military action hostile to its interests.

Clandestine Strategic Attack – A Tier One nuclear power will have the capacity to divert or sell nuclear weapons to a second party that may be a non-state actor. This could be part of a strategy to threaten the United States and/or its key allies with an unconventional means of delivery.

Militarily Operational or Tier Two Capabilities: Multi-Salvo Capability

As the arsenal expands to several tens of weapons, the regional power will be able to consider a wider range of options. At present, both India and Pakistan have nuclear arsenals that provide for limited nuclear use. Given the scale of its investment in fissile material production infrastructure, it is likely that Iran aspires to a similar capability. Including the options available to a Tier One power, these options include:

Counter-Reconnaissance Operations – The regional power could use a small number of nuclear weapons to attack US and allied reconnaissance satellites operating in LEO. Alternatively, the regional power might attack US reconnaissance aircraft in flight with several high altitude detonations to generate area EMP effects.

Theater Military Use – The regional power could use a small number of weapons, say a dozen, to support military operations designed to cripple a US expeditionary capability. Targets attacked could include major airfields, ports, logistic sites or key elements of a US forcible entry operation.

Strategic Retaliation – The regional power will hold a significant portion of its nuclear arsenal in a survivable reserve to be used in retaliation against the major cities of the neighboring states in event of US escalation with or without the use of nuclear weapons. Aside from the more expensive and technologically challenging traditional means of delivery such as an ICBM, a second Tier state may develop a wide range of less conventional delivery systems.

Tier Two Modernization and Response Options

Regional opponents with a Tier Two capacity will take further measures to deny a US counter-nuclear campaign capability capacity. The United States should assume that its regional adversaries will deliberately take advantage of weaknesses in US capabilities. Asymmetric responses will be the norm for these outgunned and outspent powers. Some of the major initiatives will likely include:

Defeating Allied Active Defenses by Developing Maneuvering Re-Entry Vehicles – Although initially developed as a countermeasure to defeat national missile defense systems during the Cold War, the concept of the maneuvering re-entry vehicle (MaRV) has matured as a means of providing ballistic missiles with precision accuracy. The Pershing II, with a radar-guided MaRV, was the Cold War generation of this technology. Now this technology takes advantage of satellite navigation technology to provide for low-cost, accurate ballistic missiles.

The United States, Russia, and China have taken the lead in refining this technology, but it will likely diffuse rapidly, either through the global arms market or by indigenous development. This could allow regional adversaries to penetrate US and allied missile defense systems.

Saturation of US and Allied Defenses Through Mass Production of LO Cruise Missiles – Similar to the technology of the turbo-fan powered Tomahawk long-range cruise missile, regional states have begun to master the mass production of a similar generation cruise missile. Relying on satellite navigation and low cost terminal seekers, a regional power may be able to produce a 21st century version of the V-1 in the thousands. Launched from ground mobile vehicles, a regional power could mount saturation attacks against defended targets that are at intermediate ranges of approximately 1,000 miles.

Defeat Missile Defenses Through Salvage Fusing and Exo-Atmospheric Countermeasures – Nuclear warheads could be "salvage fused" to detonate on impact with a kinetic energy hit to kill missile. High altitude nuclear detonations will cause either electromagnetic effects or scintillation in the ionosphere. In addition to producing "blackout," both phenomena will degrade the performance of a traditional missile defense system to deal with immediate follow-on attacks. Aside from MaRVs, the ballistic missile could be equipped with low-cost exo-atmospheric decoys that simulate the radar and EO signature of an incoming warhead.

Managing Tier One and Tier Two Powers

Strategic interactions with Tier One and Tier Two powers are likely to exhibit a greater range of dynamics, some of them potentially combustible. States with nuclear arsenals characterized by: small numbers; limited range and reliability; and unsophisticated, vulnerable, and/or imperfectly controlled nuclear arsenals are likely both to offer more plausible prospects for US military confrontation and greater chances of escalation. On the first point, the United States should not be deterred by a nation possessing a modest arsenal. First, such an arsenal will be inelastic and less employable, with weapons suited for deterrence rather than operational use. Second, US forces can be structured to continue to operate effectively in a limited nuclear environment. Additionally, US forces may be able to use counterforce capabilities to knock out or significantly degrade a poorly-situated enemy arsenal. Thus US forces will be able to act effectively against an immature nuclear power.

Conversely, however, the very plausibility of US military action against such states means that the risk of escalation is likely higher. Adversaries will be inclined to adopt variants of the Cold War approach of deliberately increasing the threat level in order to deter American action. This could include launch-on-warning or even preemptive strikes in order to "use" rather than "lose" their capabilities. Unsophisticated nuclear states might also delegate the authority to use nuclear weapons down the military chain or plan for nuclear use as part of a war plan.

Constructing a Way Ahead

To have a chance of defeating Tier One or Tier Two nuclear-armed regional opponents, future US administrations would have to commit to a major shift in defense planning and investment. The next administration would have to acknowledge that any future major regional contingency will likely involve the threat of nuclear weapons. This new reality compels a major investment in a variety of offensive and defensive counter-nuclear campaign capabilities, a much more robust C4ISR architecture, and joint theater forces designed and trained to operate under nuclear attack. Most importantly, there would have to be a major investment in the development of appropriate concepts of operation and associated training of the joint force to achieve a significant improvement in US warfighting capability. These new investments must be made with the proviso that US forces, specifically ground forces, remain prepared to conduct large scale stabilization and counterinsurgency operations. Further, any new investment will likely have to be selective since total defense resources after Iraq will be constrained by other powerful budgetary demands.[111]

Counter-Nuclear Campaign Requirements

Persistent and Survivable Reconnaissance – Today the United States has developed a wide range of persistent reconnaissance means that are not survivable in the face of advanced air defenses. Further, LEO satellite reconnaissance systems are vulnerable to non-nuclear and nuclear anti-satellite threats. What the USG needs is a variety of reconnaissance means that can fight for targeting information. This requirement can be addressed by the various combinations of persistent and/or responsive strike systems identified below.

Persistent and/or Responsive Strike – With the possible exception of the small inventory of B-2 bombers, the United States does not have a capacity to conduct a counter-force campaign that can react to near real-time intelligence on fleeting targets. This means a major investment in either the N-UCAS or a regional bomber as a high performance and survivable reconnaissance-strike platform. Alternatively, a new generation of UAS, both land and sea based, with high survivability features is needed to provide targeting information for a large inventory of rapid-response strike systems, either the HCM or a conventionally-armed ballistic missile.

Effective Munitions – Although there has been major progress in developing a new generation of munitions, especially those designed to defeat hard targets, there remains the challenge of defeating superhard and dispersed underground targets. This challenge raises the prospect that some deeply buried targets can only be attacked by a new generation of nuclear EPW. Although

---

[111] For a more complete discussion of these demands and the constraints that they and other trends will likely pose on future defense budgets, see Steven M. Kosiak, *US Defense Budget: Options and Choices for the Long Haul* (Washington, DC: Center for Strategic and Budgetary Assessments, forthcoming).

this analysis did not focus on US nuclear-use options, there remains the central question as to whether the US nuclear arsenal will play any direct warfighting roll in the types of Tier One and Tier Two challenges examined above. At present, Congress has denied funding for a R&D effort to develop such a weapon. The next administration will have to revisit this issue.

Comprehensive SEAD – To enable persistent and responsive reconnaissance-strike operations, the United States will have to maintain a powerful capacity to conduct SEAD operations against a wide range of aerospace defenses. Tier Two opponents may have the resources and military competence to acquire and deploy very robust air defense systems during the course of the next several decades. Any plausible deep reconnaissance-strike operation will require the support of a robust SEAD capability. One attraction of the rapid deep strike systems such as the HCM and conventionally armed ballistic missile is their use as "counterbattery" fires against mobile high performance SAM systems that must radiate to function.

Survivable Basing – The reconnaissance-strike and SEAD units will have be able operate from either defended ships at sea or defended bastions located some distance from the theater of operations. Naval forces provide the option of robustly defended, sea-based mobile reconnaissance strike forces, while air forces, especially with a regional bomber, will have to operate out of intermediate support bases within the theater of operation. These intermediate support bases with have to be heavily defended from air, missile, and submarine attacks with nuclear weapons. Simply larding up Guam and Diego Garcia with powerful but unprotected reconnaissance-strike assets is to invite a regional preemptive strike by either long-range precision attack or nuclear means – a replay of the Clark Air Base disaster of December 8, 1941.

New Model Combined Arms Operations

New Concepts of Operations – The Army and Marine Corps will have to develop a new joint operational concept for major contingencies that includes the prospect of combined operations under nuclear attack.

Sustaining from a Bastion – At present, only widebody airlifters such as the C-17 and C-5 can sustain brigade-sized forces over intra-theater distances. This requires that an airfield be seized by either airborne and/or amphibious assault. That forward operating airfield will be very vulnerable to a direct precision and/or nuclear attack. To avoid the use of a single fixed airfield, some form of aerial support is needed. In the near-term, it will be possible to use JPADS to provide support for maneuvering units without relying on a large airfield. Supplies can be directly delivered to the maneuver forces using the airlift fleet while it operates at medium altitude (approximately 25,000 feet). Alternative concepts discussed above include the development of a fleet of Joint Heavy Lift (JHL) VTOL cargo assault aircraft or the development of a fleet of semi-buoyant air ships. The former would require a large R&D commitment to develop a heavy lift VTOL aircraft with an IOC well after 2020. The latter will have to overcome the "giggle factor" but may represent a genuine innovation in intra-theater and strategic lift capability.

Operating from the Seabase – As noted, the Navy and Marine Corps are making an innovative investment in new seabasing technology. These sea-based forces will provide logistics support to large scale ground operations. The ships of the seabase will have to be protected from a wide range of ballistic and cruise missiles and submarine-delivered nuclear threats. One issue is whether to expand substantially the current Navy plan to deploy only one MPF (F) squadron by the middle of the next decade. Another question to be addressed is whether several more seabases should be procured by 2020 even at the expense of other high priority Navy warship programs.

Increased Role of Armored Forces – While conducting military operations under the threat of nuclear weapons, the most plausible force mix is comprised of the heavy and medium BCTs of the US Army. Both will have the inherent armor protection and mobility to more effectively survive and function after a first generation nuclear weapon detonation. Infantry BCTs, if fully equipped with MRAPs, will have survivability features similar to a Stryker brigade. To ensure that all BCTs will have their organic logistics and engineering elements equipped with armor protection, the post-Iraq War MRAP inventory could be used to replace the soft vehicles currently deployed with logistic and engineering units.

"Going to Ground" – With an organic combat engineering capability, the BCTs, and especially the division and above echelons will have to rapidly build field fortifications to reduce the consequences of nuclear weapons use. As noted, any in-theater defended bastions will have to be configured in a locally dispersed and hardened fashion.

Dispersed Operations - BCTs will have to be prepared to operate tens of miles from their neighboring BCT. The division echelon will basically be a C4ISR node for the multi-BCT operations and will have little organic fires and logistics capability other than a combat engineering capacity to "go to ground." Both the BCTs and division echelon will rely heavily on non-organic assets to provide ISR, long-range fires, combat aviation, aerospace defenses and logistics support.

Reach Back – The traditional forms of organic combat support and combat service support will have to be radically pared back to reduce the overall vulnerability of the ground maneuver units to nuclear strikes. Most of these assets, in particular all consumables like water and POL must be delivered by air from heavily protected bastions or seabases.

Reliance on Guided Munitions – All indirect fire systems whether organic to the BCT or provided by outside sources will rely heavily on guided munitions such as the *Excalibur* GPS-guided artillery shell and the GMLRS. Their use will radically reduce the logistics throughput requirements of these very logistically intensive elements of the ground maneuver units.

Refocusing the FCS Program – The central rationale for the FCS program should be to modernize the Army so that it can credibly conduct large scale ground operations against a nuclear armed opponent. In a nuclear environment, their heavier weight, lower profile, and ballistic shape, will make them much more survivable than the widely deployed MRAP family of armored vehicles.

The key enabler of the FCS program is high bandwidth communications coupled to battlefield and theater surveillance systems that provide a "first shot, first kill" capability using stand-off guided munitions. If deployed Army-wide, the enhanced C4ISR suite derived from the FCS program would enable units to operate in a nuclear shadowed environment. This C4ISR suite must be hardened against EMP effects associated with second generation high yield nuclear weapons. This calls for an EMP "tax" higher than Option B since it involves the modernization of the entire ground force.

Training for Nuclear "Shadowed" Operations – If the Army and Marine Corps return to training for conventional combined arms operations as the Iraq War winds down, they will have to develop a training syllabus that prepares field forces for nuclear attack. At a minimum, a series of exercises should be conducted to test the feasibility of Army and Marine Corps forces operating in the dispersed fashion and with the high reliance on non-organic fire, C4ISR, and logistics support as described above.

Mature Tier Three Capabilities: Assured Retaliatory Capability

The next major step in capability is for a regional power to acquire a second generation of nuclear weapons to become a mature or Tier Three nuclear power. Initially, these weapons may employ "boosting" techniques to provide for weapon yields that are a factor of two to five greater than a first generation 20 kiloton weapon. Both India and Pakistan may attempt to acquire this additional qualitative capability without the use of testing. More challenging is the development of fission-fusion thermonuclear weapons.

Tier Three powers will have far more capable and diverse systems to deliver their arsenal of 100-300 nuclear weapons. The means of delivery will include second generation solid propellant ballistic missiles, both land and sea based, first generation LO and supersonic cruise missiles; long-range, LO strike aircraft; and supersonic cruise missiles.

At present, all of the mature nuclear powers including China, France, United Kingdom, and very likely Israel have this Tier Three capability. This expanded arsenal provides additional capabilities such as:

Wide Area EMP – Thermonuclear weapons can be used at high altitude to generate very large EMP effects. This provides the regional power with a use option that uses nuclear weapons more for mass disruption rather than destruction. Further, these weapons could be used to damage US naval forces on the high seas.

Multi-Salvo Operations – Mature regional powers will have the option of conducting limited counter military strikes against a wide range of targets while holding a future opponent's major cities at risk.

Threatening City Annihilation - The major cities of any power can be threatened with complete destruction by a thermonuclear strike.

The Challenge of Mature, Tier Three Powers: Strategic interactions with mature nuclear powers are likely to tend towards stability and non-kinetic means of resolution, as serious combat is perilously prone to escalation to the nuclear level. Further, the temptation to launch disabling or surprise strikes is much less in such circumstances, in light of an opponent's secure retaliatory capability. In these relationships, US forces will face dynamics similar to those seen in the later stages of the Cold War wherein a military "victory" beyond quite limited objectives is unlikely if not impossible.

During the cold war, the US fought two protracted major contingency operations in Korea and Southeast Asia. Fearing horizontal and vertical escalation, the United States accepted stalemate in Korea and defeat in Vietnam rather than provoking a major conflict with either the Soviet Union or China.[112] Similarly, the Soviet Union accepted defeat in Afghanistan rather than pursuing war with Pakistan, which was supported by its nuclear armed patron, the United States. Conflicts between the United States and Tier Three powers will likely follow a similar pattern of limited use of military violence to shore up strategies of containment and extended deterrence rather than a decisive military victory.

---

[112] See Zhang, *Red Wings Over the Yalu*, for account of the Soviet Air Force's large-scale support of MiG-15 aerial operations against the UN air forces, specifically those of the US Air Force and Navy. During the Vietnam War China provided tens of thousands of troops inside North Vietnam to provide air defense and transportation assistance. The US consciously chose not to highlight that direct military aid vice the massive material assistance provided by both the Chinese and Soviets.

IX. CONCLUSION

In a proliferated environment, the US will need to be able to engage a variety of types of nuclear-armed opponents.

Nascent or Tier One nuclear powers: Some nations may elect to obtain nuclear capabilities strictly for deterrent purposes. Reflecting this policy, their arsenals would be tailored for secure second strike and counter-value functions in order to assure retaliation. Since their arsenals would be purely defensive and unrefined, future US dealings with these countries would likely be limited in nature. Regime change would be taken "off the table" in most scenarios. While US policymakers would have to be concerned with the prospect of escalation to the nuclear level, limited military operations themselves would not necessarily be radically altered. States that appear to have adopted minimum deterrent postures include China (until recently) and South Africa. North Korea's current capability, such as it is, probably reflects a minimum deterrent posture.

Emergent Tier Two nuclear powers: Some states may seek nuclear weapons not only for deterrent purposes but also for various operational uses. Some states may deploy lower yield, earth penetrating, neutron (anti-personnel), or other weapons tailored for straightforward warfighting purposes. Such weapons could be used against the whole spectrum of classic military targets including: deployed forces in the field, staging grounds, depots, airbases, ports, other bases, transportation networks, supply lines, and so forth.

Such weapons could also be used for indirect military purposes such as: area denial, terror and disorder inducement, and attacks against non-military economic and social targets. Israel, India, and Pakistan all appear to possess such weapons. States may also develop weapons capable of anti-technology effects to destroy or degrade an opponent's C4ISR capabilities, including EMP and anti-satellite functions.

The Shield/Sword Challenge: States such as Iran may fall somewhere in between the above two categories. For instance, a state may develop nuclear weapons primarily for strategic deterrent functions but use this limited, retaliatory arsenal to cover aggressive operations through other means, such as supporting proxies in other states. Such powers may see openings for aggression because they believe their opponents will be intimidated by the possibility of escalation to the strategic level.

Mature, Tier Three nuclear powers: A Tier Three nuclear state such as China becomes a strategic challenge similar to the Soviet Union during the Cold War. Conflict between them and the United States becomes highly stylized, with both sides showing considerable constraint. Large scale or militarily decisive war between the United States and a mature nuclear armed state becomes unworkable while in its place the strategies of deterrence, containment, and rapprochement become paramount.

Summing Up

The need to deter or defeat a nuclear adversary should drive US defense policy and programmatic decisions. Developing this facility may also deter "on the fence" powers from obtaining nuclear weapons at all. Conversely, such capabilities may incentivize Tier One nuclear powers to develop their arsenals to Tier Two or even Tier Three levels. This analysis strongly suggests the following broad initiatives should be taken by the next administration to address the emerging challenge of nuclear-armed regional powers.

Long-Range Reconnaissance-Strike – In the defense planning environment of the next decade, the US must decide if it should invest in a number of new programs to more effectively conduct deep counter-nuclear operations. Both the Air Force and Navy will have to make hard budgetary choices as to whether their current programs of record, with their heavy emphasis on tactical fighter modernization, may have be sacrificed to provide resources for these new needs.

Training for Counter-Nuclear Operations – A new generation of Rolling Sands-type exercises needs to be initiated to train the joint force in the conduct of offensive and defensive counter-nuclear operations. This effort should be given priority similar to the Navy's commitment to antisubmarine warfare exercises and training.

Operating from Defended Bases – Sufficient resources should be made available to provide key theater ground and seabases with robust active and passive defenses against at least an emerging Tier Two nuclear threat.

Preparing for Dispersed Operations – While simultaneously maintaining a capacity to conduct protracted irregular operations, both the US Army and the Marine Corps should make major investments in training at least a portion of their forces for operations in a nuclear shadowed environment. At minimum, a new generation of field exercises that emphasis the use of air support to provide deep maneuver forces with non-organic logistic support and fires should be developed.

Determining the EMP Tax – OSD and Joint Forces Command should sponsor a high level effort to ascertain how much hardening is necessary to provide next generation military capabilities that can effectively operate on a battlefield electronically affected by multiple nuclear weapon detonations.

Like the Flexible Response plan of the 1960s, the approach here advises firming US capabilities at the lower levels of the nuclear escalation ladder in order to render more flexible the otherwise brittle commitment to escalate to massive nuclear retaliation in the face of even limited nuclear use by an opponent. These recommendations do not suggest that the United States is obligated to meet enemy action correlatively or symmetrically – nuclear retaliation, massive or otherwise, may be the appropriate response – but posits that the United States should have such a capability.

Further, by taking action to prepare for a regional nuclear conflict, US military forces will thereby also obtain a very robust capacity to respond to future regional opponents armed with a large arsenal of tactical and theater range precision guided munitions.

The bottom line is that the United States will likely soon be facing a world of more nuclear powers which have obtained these weapons in order to deter, deny, or defeat US activities on behalf of its own and its allies' interests. If current policy persists, a nuclear-armed adversary may very well present the United States with a situation in which its choices are between being deterred or defeated or employing nuclear weapons. The policy laid out here is an attempt to avoid this binary choice, at least in limited nuclear war circumstances.

Appendix A: A Brief History of Nuclear Proliferation

Concerns about proliferation date back to the very beginnings of atomic power. Following the decisive demonstration of the power of the "absolute weapon" by the United States in 1945, many feared that nuclear weapons would swiftly spread and that such an eventuality would lead to catastrophe.[113] The unexpectedly early detonation of an atomic weapon by the Soviet Union in 1949 followed by the detonation of hydrogen bombs by the United States and the Soviet Union in the following half-decade and the development of an atomic weapon by the United Kingdom all seemed to point in this direction.

Though the arsenals of the two superpowers rapidly mounted, early fears of rapid and extensive proliferation proved unfounded.[114] With the exception of France, all remaining NATO countries (including Canada, which had participated in the development of the original bomb) forswore nuclear weapons, as did Japan. The major reason for this forbearance was the extended US nuclear deterrent commitment which, despite considerable anxiety about its reliability in Europe, was sufficiently credible during the Cold War to dissuade countries fully capable of developing nuclear weapons from doing so.

Furthermore, the strong US opposition to its allies' obtaining independent arsenals weighed heavily against proliferation during the early, testiest years of the Cold War when maximal American leverage coincided with center-right political dominance in Europe. By the 1970s, center-left political ascendancy and significant anti-nuclear sentiment in Europe combined with a massive American nuclear commitment to render the issue of further independent nuclear powers largely moot.

Dynamics were similar among US allies elsewhere, where the US defense commitment and opposition to allied nuclear capabilities combined to stem proliferation. Japan swore off any aspirations for nuclear weapons in its constitution and settled under the American nuclear umbrella. South Korea, defended by nuclear-capable US forces, also operated in a comparable strategic situation. Though Seoul flirted with an independent nuclear capability in the 1970s, the United States pressured it to abandon the program. Despite a more ambiguous relationship with Taiwan, a similar dynamic existed there as well, with the United States pressuring the Republic of China to abandon its independent nuclear program in the 1960s. Meanwhile, with the exception of the subsequently-regretted decision to share the technology with Beijing, the Soviet Union evinced no desire to spread nuclear capability among its Warsaw Pact allies or other client states.

---

[113] See, for instance, the Baruch Plan of 1946, which stated to the United Nations that "we must elect either World Peace or World Destruction." See "The Baruch Plan," at http://www.atomicarchive.com/Docs/Deterrence/BaruchPlan.shtml.

[114] For those concerns, see John F. Kennedy's prediction that fifteen to twenty-five states would possess nuclear weapons by the 1970s. Press Conference, 21 March 1963, Public Papers of the Presidents of the United States: John F. Kennedy, 1963 (Washington, DC, US Government Printing Office, 1964), p. 280.

The nations that persisted in developing nuclear weapons programs were those that perceived themselves to be sufficiently threatened by, isolated from, and/or suspicious of the superpowers to require an arsenal. Thus France in 1960, in light of the American response to the Suez Crisis of 1956 and skepticism about US predominance, obtained an independent nuclear capability to preserve its absolute security and independence. China, dissatisfied with Soviet supremacy in the Communist bloc and jealous of its prestige, detonated its first bomb in 1964. Isolated and highly threatened Israel, meanwhile, secretly developed and fielded a nuclear arsenal during the 1960s in the face of an intense Arab threat.

South Africa, facing international isolation by the 1960s, likewise deployed nuclear weapons to stave off the elimination of the white nationalist state. Two nonaligned European states, Sweden and Switzerland, pursued nuclear programs during the 1960s, only to abandon the efforts as their defenses became increasingly integrated, as a practical matter, into NATO's. India, threatened by China and Pakistan yet outside of the Soviet fold, tested its first weapon, secretly, in 1974. Pakistan, for obvious reasons, quickly followed. Military-run Brazil and Argentina, outside of the arena of superpower confrontation, initiated programs but were dissuaded from fielding weapons. Iraq and Iran (at war with one another) as well as Libya, all felt sufficiently threatened by the United States and Israel and internationally isolated to undertake programs during the 1980s.[115]

International opinion, along with diplomatic and legal pressure also contributed to slowing the spread of nuclear weapons. The creation of the International Atomic Energy Agency in the 1950s and the "Atoms for Peace" program were early gestures in the direction of nonproliferation. Then, in the wake of the Chinese test of 1964, the United States and the USSR cooperated to establish the Nonproliferation Treaty of 1968, which effectively divided the world into five nuclear powers – the UN Security Council's Permanent 5 – and the nuclear "have nots." Further, the expense and difficulty of constructing a civil nuclear energy backbone as well as the complexity of developing nuclear weapons dampened proliferation. The growth of a nuclear "taboo" in the years following 1945 also played a role, as did international reputation and economic reasons, in dissuading countries such as Japan, Egypt, and Mexico from embarking on weapons programs.

By the early 1990s there were five declared (United States, Russia, United Kingdom, France, and China) and three undeclared nuclear powers (India, Pakistan, and Israel). South Africa publicly abandoned its arsenal early in the decade and the 1994 Agreed Framework promised to cancel North Korea's budding capability as well. Nonproliferation and counterproliferation became higher priorities for the leading powers of the world, especially the United States, which appeared willing to take drastic action to prevent at least some forms of proliferation.[116] Moreover, the end of the stand-off between the United States and the USSR, the increasing

---

[115] For some histories of these proliferation cases, see Jeffrey Richelson, *Spying on the Bomb: American Nuclear Intelligence from Nazi Germany to Iran and North Korea* (New York, NY: W.W. Norton & Co., 2006).

[116] See, for instance, the Defense Counterproliferation Initiative announced by Secretary of Defense Les Aspin in 1993. The apotheosis of forward-leaning US counterproliferation policy was reached in the 2002 National Security Strategy of the United States.

power of conventional arms, sharp cuts in nuclear forces by the great powers, and the strengthening of the nuclear "taboo" all seemed to suggest a decreasing emphasis on nuclear weapons.

Other trends, however, pointed in the opposite direction. The decline of the bipolar world meant that interstate dynamics would no longer operate within the context of the superpower rivalry. This left a number of states, especially those dissatisfied with the "New World Order" of Western dominance and globalization, bereft of either a patron/protector or a restrainer – and thus vulnerable to the isolation that had spurred states such as Israel to obtain nuclear weapons. Further, the obvious and overwhelming dominance of US conventional arms combined with the willingness to use them, as demonstrated in Operation Desert Storm, Bosnia/Kosovo, and the initial stages of Operation Iraqi Freedom, prompted possible US opponents to conclude that confronting the United States on those terms would be foolhardy.[117] The increasing emphasis placed by the United States on preventing "rogue" states from gaining nuclear weapons also may have suggested that possession would confer significant benefits against the lone superpower.[118] As in the past, nuclear weapons were seen as "the absolute weapon," the great equalizer for a small or medium power against a great one. They thus offered both the possibility of negating the massive operational American advantage in conventional arms as well as a general strategic deterrent against the United States and other adversaries.[119] Nuclear weapons also offered a highly attractive cost-to-benefit ratio, especially for countries that no longer enjoyed the benefices of the Communist Bloc. These trends made the acquisition of nuclear weapons increasingly appealing to a number of states.

The nuclear programs of North Korea and Iran both followed these patterns. North Korea, isolated and believing itself threatened by an overwhelmingly stronger United States-led alliance, broke out of the Agreed Framework and detonated a nuclear device in October 2006.[120] The North Koreans stated that they had developed nuclear weapons for the purpose of deterring a hostile United States.[121] As the Korean Central News Agency stated, referring to the DPRK's nuclear weapons capability:

> A people without [sic] reliable war deterrent are bound to meet a tragic death and the sovereignty of their country is bound to be wantonly infringed upon...The DPRK's nuclear weapons will serve as a reliable war deterrent for protecting the

---

[117] Cited in Robert Manning, "The Nuclear Age: The Next Chapter," *Foreign* Policy, Winter 1997-1998.

[118] This seems to have been part of North Korea's calculation.

[119] For a Chinese perspective on these factors, see Major General Wu Jianguo, "Nuclear Shadows on High-Tech Warfare," *China Military Science*, Winter 1995.

[120] Whether North Korea was covertly pursuing a uranium enrichment route to nuclear weapons before 2002 is a matter of dispute. Regardless, North Korea withdrew from the Nonproliferation Treaty (NPT) in 2003 and announced its possession of nuclear weapons in 2005, signaling the demise of the Agreed Framework.

[121] "DPRK FM on Its Stand to Suspend Its Participation in Six-party Talks for Indefinite Period," DPRK Statement, February 10, 2005, at http://www.kcna.co.jp/item/2005/200502/news02/11.htm#1.

supreme interests of the state and the security of the Korean nation from the US threat of aggression....[122]

Iran also appeared to be developing a nuclear capability, though of what type and precisely why remained unclear.[123] Informed assessments, however, suggested that Iran was pursuing a weapons, or virtual weapons, capability for strategic deterrent and national prestige reasons. In particular, Iran's program likely reflected a desire to deter and offset American and Israeli nuclear and conventional military dominance (as well as the historical legacy of the long and bloody war with Iraq in the 1980s). Iran may also have been pursuing the capability to enable greater strategic maneuver room in the Middle East, including sponsoring Hezbollah and other Iranian proxies. Broadly, Iran's pursuit of nuclear capabilities conformed to the pattern of proliferation by countries that, dissatisfied with the Western-dominated order, bereft of a great power protector, and concerned about American and allied conventional supremacy, sought a "great equalizer." Syria may also have begun a nuclear weapons program, probably in light of its stark conventional inferiority to its Israeli and American foes and the end of its traditional relationship with the Soviet Union.[124]

The punctuated equilibria paradigm of proliferation would suggest that the deployment of nuclear weapons by states such as North Korea and Iran would trigger neighboring countries to consider obtaining nuclear capabilities as well. Early reports indicated that this dynamic was beginning to play out in the first decade of the 21st century. North Korea's apparent success in developing a nuclear capability and Iran's continuing efforts to do so, despite fervent American opposition, deepened anxiety among neighbors, including those aligned with the United States, about their own security in a proliferated environment. Though the United States organized international coalitions to try to force the DPRK and Iran to roll back their programs, the old problem of the credibility and limits of the American extended deterrent reappeared. The increased threat of nuclear weapons in the hands of North Korea and Iran accentuated the differences in interests between allies such as Japan and Saudi Arabia on the one hand and the United States on the other. Some reporting suggested, as a consequence, that regional adversaries were considering developing nuclear capabilities themselves.

In Tokyo, for instance, senior politicians have increasingly discussed the possibility of Japan going nuclear.[125] Likewise, in October 2006, the former head of the South Korean Grand

---

[122] "DPRK Foreign Ministry Clarifies Stand on New Measure to Bolster War Deterrent," DPRK Statement, October 3, 2006, at http://www.globalsecurity.org/wmd/library/news/dprk/2006/dprk-061004-kcna01.htm.

[123] National Intelligence Estimate, "Iran: Nuclear Intentions and Capabilities," National Intelligence Council, November 2007, available at http://www.dni.gov/press_releases/20071203_release.pdf.

[124] David Sanger and Mark Mazzetti, "Israel Struck Syrian Nuclear Project, Analysts Say," *New York Times*, October 14, 2007.

[125] For instance, then-Chief Cabinet Secretary and current Prime Minister Yasuo Fukuda said in 2002 that Japan's constitution did not preclude the possession of nuclear arms, and that circumstances have "changed to the point that even revising the constitution is being talked about." He stated that, "depending upon the world situation, circumstances and public opinion could require Japan to possess nuclear weapons." See "Japan's Nuclear Program," at http://www.globalsecurity.org/wmd/world/japan/nuke.htm. For a fuller history, see Christopher W. Hughes, "North Korea's Nuclear Weapons: Implications for the Nuclear Ambitions of Japan, South Korea, and Taiwan," *Asia Policy*, January 2007, p. 83.

National Party called for an investigation of the benefits of nuclear weapons as a balance against North Korea and a nuclear Japan.[126] Iran's pursuit was likewise prompting such considerations among its neighbors in the Middle East. Saudi Arabia, for instance, may have been cooperating with Pakistan to obtain the benefits of nuclear and ballistic missile technology.[127] Egyptian President Mubarak likewise hinted in early 2007 that if Iran fielded nuclear weapons, Egypt would as well.[128] Egypt was also discussing the development of a civil nuclear program, a capability that could be turned to weapons production if so desired. Turkey may consider a nuclear weapon option if relations with its NATO partners, specifically the United States, deteriorates over the post-Iraq War disposition of a potential emerging Kurdistan or the future spread of nuclear weapons in the Greater Middle East.[129]

In 2007, nuclear weapons remained in the hands of only a few countries.[130] However, trends suggest that the number could rise over the first decades of the 21st century. The manifest failure of international opinion or international governance mechanisms, such as the IAEA and the UN Security Council, to prevent determined countries from developing weapons capabilities was becoming increasingly clear. More importantly, the continuing allure of "the absolute weapon" as a relatively inexpensive deterrent, possible tool of intimidation, and potential equalizer was clearly appealing to those countries opposed to the US and its allies' predominance.

The stark US advantage in conventional capabilities combined with the clear willingness to use it and the intense, albeit negative, value the United States has placed on nuclear possession has stimulated adversarial states to look to the nuclear solution. For these nations, nuclear weapons offer some combination of guaranteed security and leverage against the United States and against their regional rivals.[131] In turn, these countries' neighbors will likely have to consider whether they, too, would benefit from possessing nuclear weapons. A major variable appears to be the

---

[126] Hughes, "North Korea's Nuclear Weapons: Implications for the Nuclear Ambitions of Japan, South Korea, and Taiwan."

[127] See Senate Foreign Relations Committee, *Chain Reaction: Avoiding a Nuclear Arms Race in the Middle East* (Washington, DC: GPO, Sen. Rpt 110-34, 2008), p. ix; and "Saudi Arabia Special Weapons," at http://www.globalsecurity.org/wmd/world/saudi/index.html.

[128] See Mubarak's statements at the January 2007 summit with Israeli Prime Minister Ehud Olmert. Mubarak stated: "We don't want nuclear arms in the area but we are obligated to defend ourselves. We will have to have the appropriate weapons. It is irrational that we sit and watch from the sidelines when we might be attacked at any moment." Also: "We don't want nuclear weapons. But since they appear highly present in the area, we must defend ourselves." Roee Nahmias, "Mubarak Hints: We'll Develop Nukes," *ynetnews.com*, at http://www.ynetnews.com/articles/0,7340,L-3348600,00.html.

[129] For a full discussion the impact of a nuclear-armed Iran on Saudi, Turkish, and Egyptian proliferation potential, see Senate Foreign Relations Committee, *Chain Reaction: Avoiding a Nuclear Arms Race in the Middle East*.

[130] Only the United States, Russia, China, Great Britain, France, India, Pakistan, and North Korea have declared or tested nuclear weapons. Israel is likely an undeclared nuclear power and Iran is likely developing nuclear weapons as well.

[131] It is for this reason alone that many states will view the renewed call by retired military and civilian leaders in the "West" for nuclear abolition with great suspicion. They will fear that a successful global nuclear abolition campaign will make the world safe for the US to use its superiority in IT-intensive forms of combined arms warfare against their national interests.

credibility and nature of the extended deterrence commitments of the United States to threatened partner nations such as Saudi Arabia, Japan, and the Republic of Korea.

Appendix B: History of Nuclear Operations

The detonation of the first atomic bombs ushered in a revolution in military affairs. Before the unveiling of the absolute weapon, the logic of industrial era warfare pointed in the direction of massing force and firepower to achieve overwhelming mastery over one's enemy. World War II carried this logic to its fullest conclusion, as fleets of tanks, airplanes, and ships delivered maximal firepower against the full range of enemy targets. Mass and power, with allowance for mobility, were mutually reinforcing.

The introduction of the atomic bomb – and especially fusion weapons several years later – decisively broke the connection between mass and firepower. A single multi-kiloton bomb could now deliver vastly more destruction than the 1,500-strong bomber fleets that had struck German cities during the war. In theory, as the arsenals of the United States and the USSR grew, atomic weapons in all their power could make any and all strategic efforts effectively pointless. Thenceforth, all military operations would have to take place under the shadow of this reality. The history of military affairs since that point has been in key respects an effort to understand how to fight, with what weapons, and with what limitations under the nuclear shadow.

In the United States, the attempt to incorporate the bomb into military planning proved frustrating and difficult. In the early years of the Truman Administration, atomic weapons were few in number and ill-suited to operational, rather than political-strategic, use. For political reasons, the Truman Administration elected not to employ nuclear arms during the Korean War and more or less stumbled into a limited war framework. The Eisenhower Administration, however, led by the President himself, took a very different tack, believing that nuclear weapons were usable, both as sources of leverage and at the operational level.

Eisenhower used nuclear threats to bring the Korean conflict to a close, issued a commanding directive to incorporate nuclear weapons into war planning, and generally adopted a policy of "massive retaliation" (presumptively strategic-nuclear) against Communist aggression.[132] In light of this policy, the Administration focused the much reduced post-Korean War defense budgets on the Air Force and Strategic Air Command, which was tasked with the destruction of the Communist heartlands in the event of war. Faced with this shift in emphasis, the other services, and especially the Army, struggled to adopt their missions and forces to a nuclear battlefield and much more austere budgets.[133]

The Army's effort in the 1950s to adapt to the nuclear battlefield was one of the most extended experiments in rendering ground forces operable in a nuclear environment. In light of major cuts in its manpower and funding, the Army developed and then attempted to field forces prepared to

---

[132] It is a matter of historical dispute whether Eisenhower really believed nuclear weapons were so usable and whether he actually intended to follow through on whatever threats "Massive Retaliation" was taken to imply. For our purposes here, however, it is clear from the record that military planning and force structuring reflected a clear readiness to employ nuclear weapons in the event of war.

[133] See Bacevich, *The Pentomic Era*; and Midgely, *Deadly Illusions*, for histories of this era.

fight with and against nuclear weapons. The most prominent element of this reform was the development of the "Pentomic Division," a restructuring of the traditional infantry divisional format left over from World War II. Guided by the themes of "flexibility, mobility, and dispersion" in response to the devastating power of nuclear attacks, the Pentomic Division was expected to be able both to employ and survive strikes by nuclear weapons. The division would be able to compensate for the necessity of dispersing to survive by using tactical nuclear weapons such as the "Honest John," "Corporal," and "Davy Crockett." These asymmetrically powerful systems would solve the problem of how light, lean, and dispersed units could bring to bear the firepower necessary for victory on the modern battlefield without relying on massed and logistically intensive conventional artillery fires.[134]

Despite the Pentomic concept's alleged promise, the effort was overwhelmingly judged a failure. Beyond the organizational and structural difficulties the effort confronted, the simple reality was that the Pentomic Division could not credibly survive and fight the nation's wars on a nuclear battlefield.[135] Once nuclear weapons, even tactical-strength ones, were employed on any significant scale, the Pentomic Division would be incapacitated, if not entirely destroyed. Soon the Army shifted toward an armored and mechanized force, the Reorganization of Army Divisions (ROAD) concept that emerged in the early 1960s. In fact, this division design became the model for the Army's heavy divisions throughout the Cold War era.[136]

The failure of the Pentomic Division, combined with a general movement in the United States against the policy of "Massive Retaliation" (driven in part by the growing and increasingly sophisticated Soviet atomic arsenal), brought the most serious experiment in nuclear warfighting to a close. The Kennedy Administration's strategy of "Flexible Response," endorsed by NATO in 1967, called for the services, and especially the Army, once again to focus on non-nuclear warfighting in order credibly to deter the Soviets at all levels of strategic interaction and to give the President the fullest range of options to counteract aggression. The war in Vietnam further concentrated Army attention away from nuclear warfighting. By the 1970s, parity, if not perceived Soviet superiority, at the strategic level channeled US and NATO efforts towards fighting a purely conventional war.

The guided weapons revolution which began to be recognized in the 1970s and came to fruition in the decades following further reduced the salience of nuclear weapons to US and NATO war-planning. This IT-intensive RMA capitalized on tremendous advances in computing, electronics, and related technologies to develop precision guided weapons and systems capable of improving the effectiveness of conventional forces against other conventional forces by an order of magnitude. Policy decisions by the Carter and Reagan Administrations, including the critical

---

[134] Bacevich, *The Pentomic Era*; and Midgely, *Deadly Illusions*.

[135] As noted in main text, the Pentomic Infantry Division design was seriously deficient in its own terms. Armored personnel carriers were held at the division level to be used by the five "battlegroups" or large battalions on demand. This is similar to the current structure of the Marine Corps with its amphibious assault vehicles, the AAV-7, held at the division level.

[136] Bacevich, *The Pentomic Era*; and Midgely, *Deadly Illusions*.

decision against the Neutron Bomb, reflected a continuing political preference for conventional over nuclear options.

Thus, although NATO retained its historical threat to escalate a war with the Warsaw Pact to the strategic nuclear level, in the 1980s for the first time in its rivalry with the Pact, the Alliance realistically prepared to defeat Communist forces in purely conventional combat. While this policy was never put to the test in the crucible of superpower combat in Europe, it did appear to succeed with flying colors during Operation Desert Storm and, at least in certain respects, during the wars in the former Yugoslavia and in Operation Iraqi Freedom. So dominant and effective were American conventional forces that many, including military personnel, judged nuclear weapons to be no longer usable, given the strength of modern conventional forces, the enduring nuclear taboo, and the "madness" of nuclear weapons. Indeed, by the first decades of the 21st century, nuclear warfighting seemed a distant problem for the US military, even an anachronism.

THE SOVIET EXPERIENCE

Though the history of Soviet and Warsaw Pact planning for nuclear operations remains in large part obscure, a general picture has emerged. During the Cold War, Soviet theorizing on warfare tended to remain more faithful to the "Clausewitzean" notion that warfare, even in the nuclear age, remained the servant of political purposes; therefore nuclear weapons could be used in the service of rational objectives. The Soviet saw this as the only interpretation of the introduction of nuclear weapons commensurate with Marxist-Leninist principles.[137]

Therefore, the Soviets, though they became increasingly interested in the possibility of solely conventional warfare later in the Cold War, appear to have never deviated from the position that nuclear weapons could and likely would have been a part of any conflict with NATO, and that military operations could be conducted successfully in such an environment. This point seems to have been confirmed by the opening of satellite nation archives in the 1990s, which showed that the Warsaw Pact had realistically planned and trained to employ limited nuclear strikes against NATO, at least in some scenarios.[138]

---

[137] For more on Soviet military doctrine, see W.C. Frank, Jr. and P.S. Gillette, eds. *Soviet Military Doctrine from Lenin to Gorbachev, 1915-1991* (Westport, CT: Greenword Press, 1992); David M. Glantz, *The Military Strategy of the Soviet Union: A History* (London: Frank Cass, 1992); and R. Craig Nation, *Black Earth: Red Star: A History of Soviet Security Policy, 1917-1991* (Ithaca, NY: Cornell University Press, 1992).

[138] For example, the Warsaw Pact War Plan of 1964 detailed the Soviets plan for the extensive use of nuclear weapons in the event of war with NATO. Petr Luňák, "The Warsaw Pact War Plan of 1964," available at http://www.php.isn.ethz.ch/collections/colltopic.cfm?lng=en&id=15365. See "Strategic Operations of the Nuclear Forces," available at http://se2.isn.ch/serviceengine/FileContent?serviceID=PHP&fileid=D3861DEF-4DF1-76D8-7707-1E873AB3793C&lng=en, for a Soviet Study of the Conduct of War in Nuclear Conditions, especially the section on the "Operations of Ground Forces." See also Lothar Ruhl, "Offensive Defence in the Warsaw Pact," *Survival*, September/October 1991, p. 446; and Beatrice Heuser, "Warsaw Pact Military Doctrines in the 1970s and 1980s: Findings in the East German Archives," *Comparative Strategy*, 1993, pp. 437-457. For earlier assessments, see William Odom, "The Soviet Approach to Nuclear Weapons: A Historical Review" *Annals of the American Academy of Political and Social Science*, September 1983, pp. 117-135; and Stephen M. Meyer, *Soviet Theatre Nuclear Forces: Part I: Development of Doctrine and Objectives* (London: International Institute for Strategic Studies, 1984).

The Soviets, because of their strategic inferiority during the 1940s and 1950s, initially emphasized conventional forces. Then, after their own nuclearization, they focused on the necessarily general character of any nuclear exchange, a policy intended to deter the Americans from using their stark numerical and qualitative advantage in nuclear weapons. Under Khrushchev and with the maturation of the Soviet arsenal, however, the Pact began to emphasize the incorporation of nuclear weapons into real warfighting.[139] Greater confidence in the relative strength of the Soviet arsenal and growing realization of the unpredictable and catastrophic consequences of "going nuclear," however, enabled (or forced) Pact planners to explore limited nuclear and even purely conventional options.[140] Still, until the very end of the Warsaw Pact's existence, war planners prepared for victory through nuclear exchange.

Whether Soviet and Warsaw Pact planning was realistic and whether Pact forces actually could have operated to the extent expected under nuclear conditions remain disputed. What does seem clear is that many of the same lessons that emerged from the American experience also came out of the Soviet one.

---

[139] See, for instance, a speech by Soviet Defense Minister Marshall Malinovskii cited in V. Mastny and M. Byrne, eds., *A Cardboard Castle? An Inside History of the Warsaw Pact, 1955-1991* (Budapest: Central European University Press, 2005), pp. 120-1. As the editors describe, "absurd though [these nuclear war plans] were, the scenarios were nevertheless meant seriously. There cannot be a doubt that the plans that had been drawn were intended to be put into effect if war came, whatever the chances that they could actually be successfully implemented." The Soviet deployment to Cuba during the Cuban Missile Crisis is an example of the nuclear warfighting orientation of the Red Army by the early 1960s. By that time much of the Soviet military agreed that nuclear arms were highly efficient battlefield and operational strike systems and were needed to facilitate the very high rates of advance required in the event of a NATO/WP conflict in Europe.

[140] Ibid., pp. 81, 120. By the early 1980s, the Soviet military leadership had become less and less enamored with the prospects of nuclear operations and were shifting their emphasis toward high technology, non-nuclear operations. Marshall Ogarkov and his supporters within the Red Army became convinced that a new "revolution in military technical affairs" was about to overtake the Cold War arms competition. This indicated the rising appreciation in the Soviet military of the United States' development of an increasingly IT-intensive form of warfare, dominated by the use of precision-guided munitions. This concept emerged in the US as "transformation" or network centric warfare.

APPENDIX C: NUCLEAR WEAPONS EFFECTS

Nuclear weapons have five primary categories of effects: blast, prompt ionizing radiation, thermal radiation, long-term ionizing radiation (fallout), and electromagnetic pulse (EMP). Each is described below.[141]

BLAST EFFECTS

Blast effects are essentially instantaneous and the most destructive for physical structures and people. Immediately following detonation, a high-pressure shock wave develops and moves outward from the point of the detonation. Similar to conventional explosives, the majority of the destructive power of a nuclear weapon results from the blast effects. The effects on infrastructure and services can be severe and long lasting, particularly within the 5 psi ring. Disruptions will also be significant within the 2 psi ring, but can be overcome more rapidly. Terrain features can provide some protection from this effect (for example, by being on the opposite side of a hill) or can intensify it (by being on the facing side of a hill). (See Table 1 for the effects of a 10 kiloton nuclear detonation.)

PROMPT IONIZING RADIATION EFFECTS

Prompt radiation is also essentially instantaneous. For doses of 600 to 1,000 rem, it is estimated that 90-100% of casualties, respectively, will be fatal even with medical treatment. For doses of 300 rem, it is estimated that 10 percent of casualties will be fatal with treatment. Below 200 rem there will be no fatalities from prompt radiation but people will suffer symptoms of radiation sickness. (See Table 1)

THERMAL RADIATION EFFECTS

Thermal radiation (heat) will cause burns to skin directly exposed to heat from the detonation. Secondary effects of nuclear weapons include fires that are started by thermal radiation and by gas and electrical lines that are damaged by the blast. (See Table 1)

---

[141] The following summary of nuclear weapon effects uses the analysis of David Howell at the RAND Corporation; Glasstone and Dolan, *The Effects of Nuclear Weapons*; and other sources.

Table 1: Parameters of a 10-Kiloton Nuclear Detonation

Overpressure (peak winds)					
12 psi (330 mph)	0.6 km (1,900 ft)	98	2	0	Most buildings collapsed, houses rubble
5 psi (160 mph)	1.0 km (3,200 ft)	50	40	10	Heavy construction severely damaged, houses destroyed
2 psi (70 mph)	1.8 km (5,800 ft)	5	45	50	Walls of buildings blown away, severe damage to residences
1 psi (35 mph)	2.5 km (8,400 ft)	0	10	90	Damage to structures, people injured by flying debris
Prompt radiation[142]					
1000 rem	1.1 km (3,700 ft)	100			Palliative treatment only
600 rem	1.2 km (4,000 ft)	90	10	0	10 percent chance of survival with treatment
300 rem	1.3 km (4,500 ft)	10	90	0	90 percent survival with treatment
200 rem	1.4 km (4,800 ft)	0	100	0	Symptoms of radiation sickness, will recover without treatment
100 rem	1.5 km (5,000 ft)	0	100	0	Mild symptoms, little effect
50 rem	1.7 km (5,600 ft)	0	0	100	No noticeable effects
10 rem	2.1 km (6,900 ft)	0	0	100	
Thermal radiation					
3rd degree burns	2.1 km (6,900 ft)				
2nd degree burns	2.6 km (8,500 ft)				
1st degree burns	3.5 km (11,600 ft)				

---

[142] Fatality rates assume adequate medical treatment is available.

LOCAL RADIOACTIVE FALLOUT EFFECTS

Radioactive fallout starts to deposit 10-15 minutes after the detonation in the area around ground zero and continues to deposit locally over the next 24-48 hours from a cloud determined by wind and other meteorological and geographic factors. The direction, intensity and dispersal of the fallout are highly dependent on local conditions and cannot be predicted.

Radiation has a cumulative effect on the body. Doses above 600 rem accumulated over a week or less are 90 percent fatal with treatment; 300 rem is 10 percent fatal with treatment, 200 rem is non-fatal even without treatment, 50 rem causes no acute symptoms.

Exposures of 50 rem or lower will increase the likelihood of long-term cancer death, however only by 0.4 to 2.5 percent. Even non-fatal radiation effects can increase chances of death from burns, other traumas, and infection. The effects of long-term exposure to low levels of fallout radiation are primarily an increased incidence of cancer and genetic mutations, although the magnitude of the effect is a matter of debate. The standard approach used by US regulatory agencies is to assume that the effect is linear and to use the factor of 1 additional cancer death for every 2,500 people exposed to a cumulative dose of 1 rem, whether it was absorbed in 1 month or 10 years. Areas within the fallout zone are likely to remain uninhabitable for years.

ELECTROMAGNETIC PULSE

Another effect of nuclear explosions is electromagnetic pulse, a phenomenon that is different depending on whether the detonation takes place at or near ground level or at high altitude.

When a nuclear detonation takes place at or near ground level, a direct electromagnetic pulse from the explosion will damage components attached to large antennas and power lines within a limited range from the explosion, depending on yield. For example, the EMP effects for a 10-20 kiloton ground burst may only extend a few miles from the site of the detonation. A ground or low-altitude burst can also destroy computers within a limited range. It is also likely to affect communications networks, but the duration of the disruption is unknown.

HEMP effects are less well understood, but could be substantial depending on the height of burst and weapon yield. In considering HEMP effects, it must be kept in mind that assessments in this realm of nuclear weapons effects are derivative of a very limited number of high-altitude tests and that there is considerable debate and uncertainty about the physical phenomena involved and the magnitude of such effects. The estimates of damage provided in the annex that follows are based on a recent review of this phenomenon done by RAND.

Introduction

In 1962 during a series of high-altitude nuclear explosion (HANE) or high-altitude nuclear detonation tests in Johnson Atoll, the US observed a variety of high-altitude electromagnetic pulse phenomena – leading to the recognition that HEMP can potentially cause debilitating damage to electronic systems within line-of-sight of the detonation. Such damage can potentially be very widespread: the Johnson Atoll tests, for instance, caused street lights to fail, burglar alarms to go off, and power line circuit breakers to open 800 miles away in Hawaii.

This primer will first explain the physical mechanics of HEMP, then describe which systems are most likely to be affected, and conclude with a list of some preventive measures.

Physical Mechanics

All nuclear detonations at heights greater than 40 km above the earth produce HEMP effects. The specific HEMP fields are dependent on the design and yield of the nuclear device. HEMP occurs in three phases—early, intermediate, and late. While the overall pulse is broadband, each phase covers a distinct energy range (see Figure 1).

A high-altitude nuclear explosion produces a large quantity of gamma rays, which travel radially away from the blast and, given the thin atmosphere, are able to propagate a significant distance. As those gamma rays traveling towards the earth reach the higher density atmosphere (20-40 km above the earth's surface), they collide with air molecules, producing free (Compton) electrons. These Compton electrons, in turn, interact with the earth's magnetic field, producing the early-time (E1) pulse that radiates towards the ground at the speed of light (see Figure 2). The early-time pulse lasts only a microsecond before completely dissipating. During this microsecond, however, the pulse produces electromagnetic fields of up to 50,000 (or more for specially designed weapons) volts per meter (V/m) in strength.

The intermediate (E2) pulse is generated by gamma rays delayed by scattering collisions or those produced by inelastic neutron interactions with air molecules. This phase is significantly less intense (producing hundreds, rather than thousands of V/m) than the first. It is similar to lightning.

The late phase (E3) is also less intense but lasts much longer. During this phase, fields of up to 1 kilovolt can last for up to several minutes. This can be particularly worrisome for long-line telecommunications and power transmission systems which act as antennas to couple this E3 energy in ways that might be very destructive.

Figure 1: Energy Plot of HEMP Phases

Figure 2: Generation of HEMP by a High-Altitude Nuclear Explosion

Effects of Hemp

HEMP interferes with electronic equipment by causing physical damage or temporary impairment to electric components. The precise effects of HEMP to a particular electronic system are often difficult to predict. A multitude of cables, pins, connectors, and devices in real hardware makes calculation complex, and the degree of damage to equipment further depends on its operational mode, shape, size, and surrounding environment. The Limited Test Ban Treaty has precluded high-altitude tests since 1963, and, as a result, experimental data is limited to data from EMP simulators.

HEMP has the potential to damage all unprotected electronic systems within line-of-sight of the explosion. Depending on the height of blast, the effects can span entire continents. A 500 km (~300 miles) nuclear explosion over Nebraska, for example, results in an EMP that covers the entire continental US (see Figure 3). To have massive disruptive effects on IT-intensive civilian and military systems, the nuclear weapon used would either have to be a lower yield fission weapon optimized to create a HEMP effect or a larger thermonuclear device, both classes of weapons available to either late Tier Two or Tier Three nuclear-armed powers.

Figure 3: Area Affected by HEMP by Height of Burst (US)

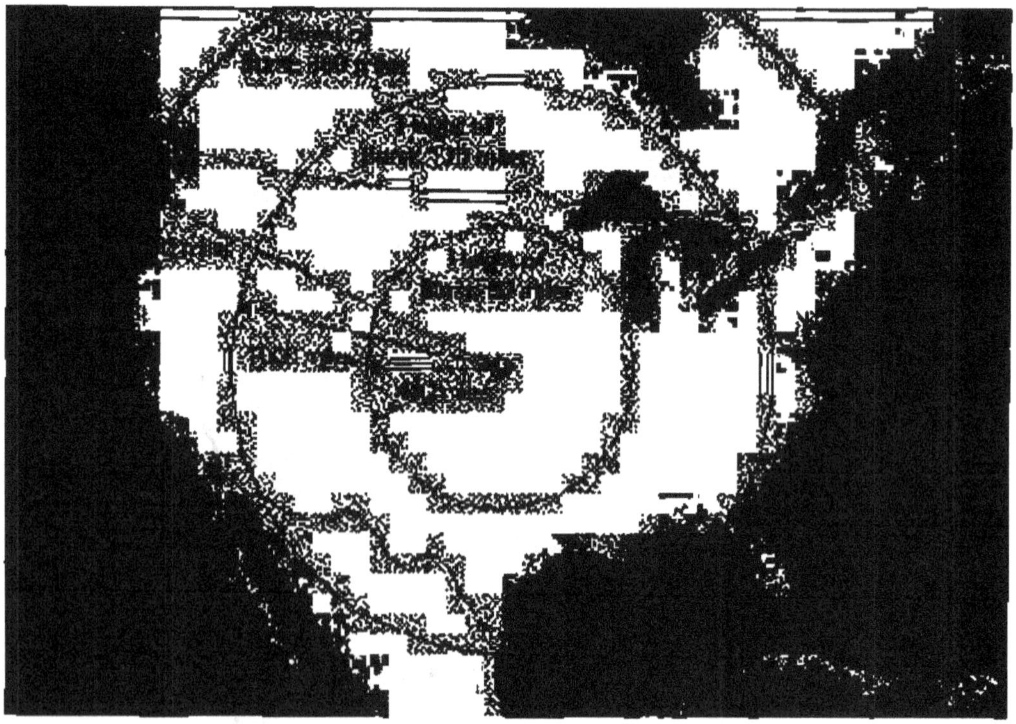

In addition, lower altitude nuclear explosions can create a more severe EMP of more limited geographic extent (see Figure 4).

The pulse from a HAND also produces different ground intensities depending on geographic relation to the burst. The contours of the intensities form a "smile" pattern as seen in Figure 4.

Figure 4: HEMP Ground Intensity for a 300 km Altitude One Megaton Burst

Effects on Ground Systems

HEMP affects electrically conductive material by coupling with conductors matched to these frequencies. When this coupling occurs, the HEMP induces high voltages and currents in the conductors, which in turn, may cause physical damage to the equipment—such as shorting or burnout of capacitors and transistors—or results in operational upset by triggering automatic shut-down or opening circuit breakers.

Because the three different phases of HEMP each contain different frequencies, each phase will couple with different equipment. The E1 phase which produces the most intense fields, couples well to antennas, conducting lines of any length, and can also affect equipment in buildings through apertures. E2 energy couples well to overhead and buried conductive lines approximately 1 km in length, while E3 couples well to conductors longer than 10 km, such as power and communications lines.

E1 phase damage tends to be destructive to small electronics such as microprocessors and other equipment using Large Scale Integrated (LSI) components, whereas E2 is normally less damaging due to electromagnetic interference (EMI) protection (similar to lightning) and may simply disrupt operations. E3, however, has the potential to damage high value power system components such as transformers, particularly if the E1 pulse has disabled protective devices and Supervisory Control and Data Acquisition (SCADA) systems. In general, HEMP can disable computer networks and critical infrastructure supporting power and communications. In addition, HEMP can penetrate the earth several feet to affect underground lines and metal water pipes, though damage is usually less severe. Unhardened airplanes are also susceptible. While fiber optic cables are not susceptible, the switches and controls that use microelectronics in conjunction with fiber optics can be damaged. There is considerable controversy about the vulnerability of many contemporary IT systems. For example, modern automobiles rely heavily on electronic systems for their functionality. These high performance solid state systems are inherently more susceptible to EMP due to their density and low voltage features. On the hand, these and many military systems have been designed to operate in the face of electromagnetic interference and are relatively "hard" to EMP effects especially those caused by a low-yield first generation fission weapon.

Hemp Effects on Other Systems

High-altitude nuclear explosions can also damage satellites or interfere with remote sensing and satellite-ground communications. In particular, a HAND would produce both an EMP and "pump-up" the van Allen radiation belt around the earth. This radiation could disable all non-hardened satellites in low earth orbit within a few days to months. Analysis by DTRA suggests that a single 20-kiloton nuclear detonation at the appropriate altitude will induce the electronic equivalent of accelerated Alzheimer's disease for most satellites operating in LEO. While most military satellites are hardened against radiation effects (both natural and threat-related), almost no commercial satellites are hardened beyond normal background levels, leaving them vulnerable if nuclear weapons pump-up the radiation belts. Satellites operating at MEO and GEO will not be affected by this phenomenon. Furthermore, satellites operating at those altitudes are much more robust since they are designed to survive bursts of solar radiation outside of the protective cover of the van Allen belt.

Even if satellites survive a HAND (or are out of range) in the near-term, the explosion itself may interrupt satellite-ground communications and satellite sensing for tens of hours. The blast creates large, highly ionized regions in the upper atmosphere. Depending on the weapons yield, height-of-burst, and location of the explosion, the ionized regions will interfere with electromagnetic waves crucial for satellite communications, satellite remote sensing, or earth observation.

Preventative Measures

While strategic nuclear forces and strategic C^3 systems are hardened against EMP, current conventional forces lack comprehensive hardening against EMP, and commercial systems possess almost no protection against EMP attack. Further, tactical unit training does not currently include HEMP scenarios.

It is generally estimated that designing hardened equipment can add 1-5% to their cost. On the other hand, retrofitting HEMP hardening can be much more costly. There are some straightforward, more immediate steps—which interfere with operations to varying degrees—that can be taken to mitigate the effects of HEMP on unhardened equipment in-theater:

- Install surge protectors specifically designed for EMP (lightning surge protectors are inadequate), which are inexpensive and effective at clipping the pulses associated with EMP on power and signal lines.

- Use radios operating at UHF and Super High Frequency (SHF), which are less susceptible to EMP.

Keep cable and wire runs as short as possible and avoid loops, which pick up more EMP than straight runs.

Keep cables and wires on the ground where possible. Elevating cables and wires increases the voltages and currents generated by EMP. Cables can also be shielded under metallic conduits.

Critical systems can be "globally" shielded under a cage of reasonably thin and conductive material (such as aluminum, copper, or steel) that is free from apertures and gaps.

Shutting-off or disconnecting equipment from cables may protect those systems with susceptible semiconductor junctions (such as the PC peripheral line interface).

Good grounding can help shunt the EMP away from equipment.

Table 2 provides a summary assessment of the degree of susceptibility to HEMP of various classes of equipment.[143]

---

[143] One of the unresolved mysteries of the Cold War is whether the Soviet military consciously decided to stay with small vacuum tube technology because they were convinced that any NATO/WP war would involve the mass use of nuclear weapons. The conventional wisdom of the 1960s is that the US got the jump on the Soviets in the development of Large Scale Integrated (LSI) and Very Large Scale Integrated (VLSI) systems because of a major lag in Soviet electronic science and industrial capacities. It is possible, though not confirmed, that the Soviet military-scientific leadership showed less interest in solid state electronics out of concern about their acute vulnerability to EMP.

Table 2: Degrees of Susceptibility to HEMP

Most Susceptible	Computers
	Integrated circuits
	Alarm systems
	Electronic sensors
	Broadband radios
	Systems employing transistors or semiconductor rectifiers: • Semiconductor components terminating long cable runs • Life-support system controls • Transistorized receivers and transmitters • Power system controls and communications links
Less Susceptible	Vacuum-tube equipment that does not include semiconductor rectifiers: • Transmitters • Receivers • Power supplies
	Equipment employing low-current switches, relays, meters • Power system control panels • Panel indicators and status boards • Process controls
	Hazardous equipment containing: • Detonators • Explosive mixtures • Rocket fuels • Pyrotechnical devices
	SHF/UHF radios
Least Susceptible	• Motors, transformers, circuit breakers, etc designed for high-voltage application
	Radios with short antennas

USING SCALING LAWS TO ESTIMATE NUCLEAR WEAPONS EFFECTS

Many nuclear effects tools show calculations for a one kiloton yield. The reason for this is that it is quite easy to use scaling laws to estimate parameters for other yields. To calculate distance parameters the scaling law is as follows:

$$\frac{D}{D_1} = \left(\frac{W}{W_1}\right)^{1/3}$$

Where:

D = Distance for estimated data point
W = Yield for estimated data point
D_1 = Distance for 1-kiloton yield
W_1 = 1 kiloton

By setting W_1 = 1 kiloton, the scaling equation becomes:

$$D = D_1 \times W^{1/3}$$

These scaling laws are presented by Glasstone and Dolan. They claim "[f]ull-scale tests have shown this relationship between distance and energy yield to hold for yields up to (and including) the megaton range."[144]

To estimate the optimum height of burst for a range of peak overpressures one can use a variant of the Glasstone and Dolan figures.

---

[144] Glasstone and Dolan, *The Effects of Nuclear Weapons*, pp. 100–101.

Figure 5: Optimum Height of Burst for a One Kiloton Blast as a Function of Ground Range for Various Pressures[145]

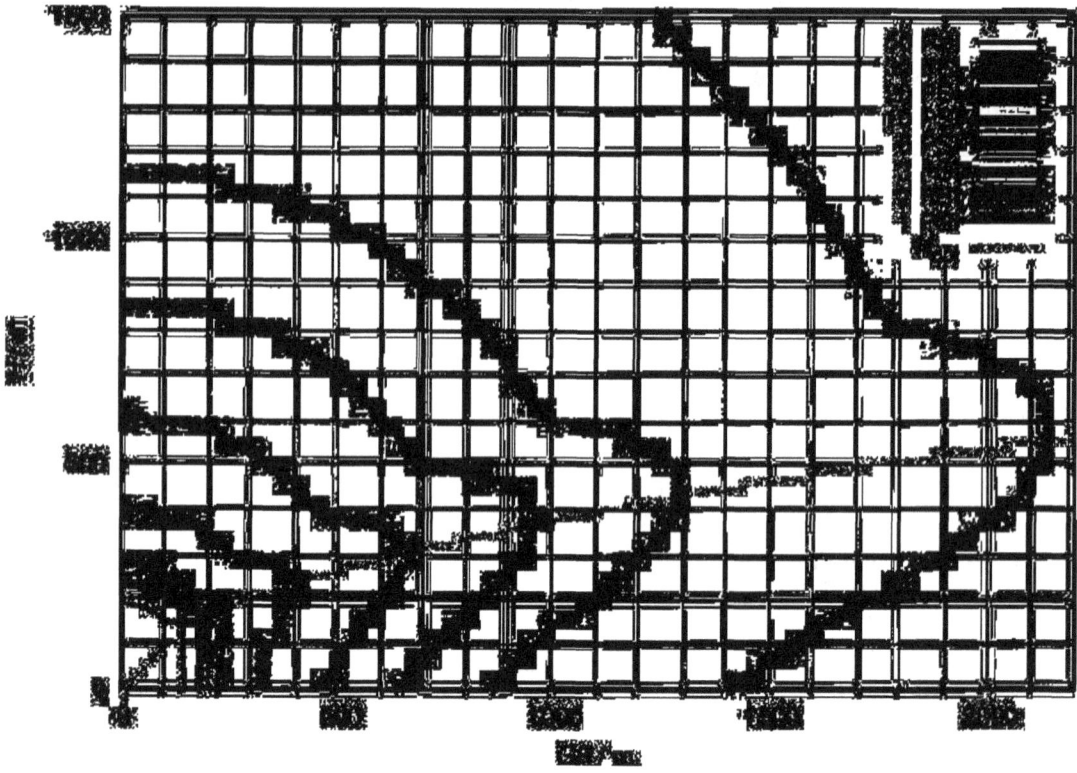

Using this graph one can determine the height of burst that maximizes/optimizes the ground range for a particular overpressure ring for a one kiloton blast. The following table shows some estimates.[146]

Table 3: Optimal Height of Burst Characteristics

Pressure (psi)	OHOB (m)	Ground Range (m)
1	550	2,150
2	430	1,300
3	370	980
5	300	650
10	250	400
20	190	250
30	160	200
50	140	150

It is important to recognize that only one of these peak overpressures can be optimized per warhead. For example, if a regional opponent detonated a warhead such that it optimized the

---

[145] Diagram available at http://en.wikipedia.org/wiki/Image:Blastcurves_1.png.

[146] David Howell, RAND Corporation.

ground range for a 50 psi peak overpressure, the ground range for all other values of overpressure will be suboptimal.

www.ingramcontent.com/pod-product-compliance
Lightning Source LLC
Chambersburg PA
CBHW080346170426
43194CB00014B/2706